Lecture Notes in Earth Sciences 44

Editors:
S. Bhattacharji, Brooklyn
G. M. Friedman, Brooklyn and Troy
H. J. Neugebauer, Bonn
A. Seilacher, Tuebingen

Deborah Anne Edwards

Turbidity Currents: Dynamics, Deposits and Reversals

With 84 Figures and 11 Tables

Springer-Verlag Berlin Heidelberg GmbH

Author

Deborah Anne Edwards
41, Mill Road
Stourport-on-Severn
Worcs, DY13 9BJ, UK

"For all Lecture Notes in Earth Sciences published till now please see final pages of the book"

ISBN 978-3-540-56123-1   ISBN 978-3-540-47475-3 (eBook)
DOI 10.1007/978-3-540-47475-3

© Springer-Verlag Berlin Heidelberg 1993
Originally published by Springer-Verlag Berlin Heidelberg New York in 1993

32/3140-543210 - Printed on acid-free paper

PREFACE

Turbidity currents have been comprehensively studied in the past although much remains unknown about both their flow characteristics and resultant sedimentary deposits. Much of this uncertainty arises from the catastrophic nature of their formation which makes them difficult to study in the environment, and has resulted in the majority of studies being experimental or theoretical.

Experiments have shown that reversals in the flow of density currents can be associated with the generation of internal solitary waves. This is in contrast to the belief held by many workers that the reversal of a turbidity current simply generates an identical flow travelling in the opposite direction.

This book arose from the need for a detailed experimental study to examine the effects and to consider the consequences of density current reversals from a variety of obstructions to their flow.

The first part of this book comprises a detailed review of literature covering the fluid dynamics and sedimentology relevant to the experimental study (chapter one). Chapter two presents the results from the comprehensive experimental programme which are discussed and compared with appropiate theoretical hypotheses. This permits the synthesis of a model for the general features of flows that result from the incidence of density currents upon obstructions to the flow. The application of this model to both modern and ancient turbidite systems is then discussed in chapter three.

This book is suitable for earth scientists with an interest in the dynamics of turbidity currents. In addition, workers from other fields such as applied maths, meteorology and engineering who have an interest in density currents and bores in practical situations may find it useful.

Thanks must go to Drs. J. Best, M. Leeder, H. Pantin, K. Pickering and B. Kneller for their help with the study and many useful discussions.

For help with the practical aspects of the project I am grateful to the following for their help: S. White, J. Mott, D. Middleton, J. Lawes and N. Woodhouse.

Leeds, November 1992   Deborah Edwards

CONTENTS

CHAPTER ONE

BACKGROUND TO THE RESEARCH AND REVIEW OF DENSITY AND TURBIDITY CURRENT FLOW

CHAPTER TWO

EXPERIMENTAL STUDY OF THE INCIDENCE OF DENSITY
CURRENTS UPON RAMPS AND OBSTACLES.

CHAPTER THREE

A MODEL FOR THE INCIDENCE OF GRAVITY CURRENTS ON OBSTRUCTIONS TO FLOW AND A DISCUSSION OF ITS APPLICABILITY TO MODERN AND ANCIENT TURBIDITE SYSTEMS

x

APPENDICES

GLOSSARY OF NOTATION

Metric units of measure are used throughout.

$a =$	Amplitude of waveform
$a_t =$	Height from base to crest of waveform
$c =$	Crestal wave velocity
$c' =$	Free surface wave velocity
$c_0 =$	$(gH)^{1/2}$
$c_0' =$	$\left(\dfrac{g\Delta\rho\, d_0'd_0}{\rho_1\, H}\right)^{1/2}$
$c_1 =$	$3c_0/2H$
$c_2 =$	$c_0H^2/6$
$C =$	Mean value of concentration
$C_f =$	Drag coefficient
$CTA =$	Constant Temperature Anemometry
$d =$	Height of cavity
$d' =$	Fluid depth beneath cavity
$d_0 =$	Flow thickness before reverse flow (bore)
$d_0' =$	Undisturbed upper flow thickness
$d_1 =$	Mean depth of reverse flow
$d_2 =$	Depth after wave within reverse flow (bore)
$d_3 =$	Depth upstream of bore
$D =$	Height of bulge at ramp
$D' =$	Grain diameter
$Fr =$	Froude number $= U/(g'h)^{1/2}$
$g =$	Acceleration due to gravity
$g' =$	$g(\rho_1-\rho_2)/\rho_1$, unless specified as otherwise in text
$h =$	Thickness over which density change occurs in a stratified fluid
$h_0 =$	Depth of denser fluid in lock prior to release
$h_1 =$	Head thickness
$h_2 =$	Body thickness of a density current
$h_3 =$	Mean thickness of mixing region behind head
$H =$	Total fluid depth

$k =$ Wave number

$L =$ Measure of horizontal wave extent

$L' =$ Magnitude of resultant vector (%)

$L_{KDV} =$ Horizontal extent of wave in the shallow water internal solitary wave theory.

$L_{BDO} =$ Horizontal extent of wave in the deep water internal solitary wave theory.

$m =$ $1/L_{KDV}$

$n =$ L_{BDO}

$n' =$ number of readings

$O\ (*) =$ To the order of **

$r =$ Boundary roughness

$t =$ Time

$u =$ Velocity past cavity along way down stream

$u =$ Mean flow velocity

$u_1 =$ Velocity in lower fluid in a 2-layer fluid system

$u_2 =$ Velocity of upper fluid in a 2-layer fluid system

$u_m =$ Velocity at the point on a density current interface where billows are initiated

$U' =$ Shear velocity

$U_h =$ Propagation velocity of the head of a density current

$U_b =$ Velocity of advance of bore

$U_t =$ Velocity in tail of density current

$R =$ Magnitude of resultant vector

$Re =$ Reynolds number $= U_h h_1/v$

$Ri =$ Richardson number $= 1/Fr^2$

$v' =$ Vertical velocity component

$V_s =$ Settling velocity of solid particles in a flow

$x =$ Horizontal co-ordinate

$x =$ Vector mean

$\Delta =$ Uniform loss of energy from head of density current

$\Delta u =$ Difference in velocity across the front of the density current

$\Delta\rho =$ $\quad$ $\rho_1 - \rho_2$

$\eta =$ $\quad$ Displacement on wave profile above quiescent water depth, $\eta(x,t)$

$\eta_t =$ $\quad$ $d\eta/dt$

$\eta_x =$ $\quad$ $d\eta/dx$

$\eta_{xx} =$ $\quad$ $d^2\eta/dx^2$

$\eta_{xxx} =$ $\quad$ $d^3\eta/dx^3$

$\lambda =$ $\quad$ Wavelength

$\theta =$ $\quad$ Shields parameter

$\rho =$ $\quad$ Density

$\rho_1 =$ $\quad$ Density of denser fluid in system

$\rho_2 =$ $\quad$ Density of less dense fluid in system

$\rho_m =$ $\quad$ $(\rho_1 + \rho_2)/2$

$\rho_s =$ $\quad$ sediment density

$\tau =$ $\quad$ Bed shear stress

$\omega =$ $\quad$ Angular frequency

$\omega_s =$ $\quad$ Grain settling velocity

$\nu =$ $\quad$ Kinematic viscosity

$\psi =$ $\quad$ $0.5d_0(\rho_2/\rho_1)$

CHAPTER ONE

BACKGROUND TO THE RESEARCH AND REVIEW OF DENSITY AND TURBIDITY CURRENT FLOW

1.1   PROJECT RATIONALE

Density currents are primarily horizontal flows that occur when a fluid of one density flows into another of a different density. They are an important phenomena in both natural and man-made situations and have been the subject of considerable research in a variety of scientific disciplines. Examples of natural density currents include the intrusion of a front of cold air under warmer air in the atmosphere, the flow of powder-snow avalanches, the entry of brackish water from an estuary into the sea, and the flow of sub-aerial pyroclastic flows and subaqueous turbidity currents. Man-made density currents are often associated with environmental pollution, with examples including the spread of an oil-slick on the sea and the release of dense gases from industry (Simpson 1987).

Turbidity currents are generated in a wide range of sedimentary environments, with turbidite deposits assuming great significance within the stratigraphic record. However, despite their clear geological importance and the considerable attention that these deposits have attracted, there has been only limited research devoted to the fluid dynamic processes of turbidity currents, their turbulent structure and their relationship to sediment transport and depositional facies. One of the major problems associated with such a study, is the impracticality of studying natural turbidity currents and as a consequence, it has meant that research has largely had to rely upon experimental and theoretical modelling.

To date, turbidity current research has concentrated upon the dynamics of the unidirectional flow currents and the consideration of the sedimentary deposits that such a flow would produce. The generation of reversed or deflected flows caused by the incidence of turbidity currents upon flow obstructions has been given limited consideration, such currents often being assumed to possess many flow characteristics similar to those of the forward turbidity current and as such, generation of similar sequences of sedimentary structures.

Recent research has shown that the flow generated by the incidence of density currents upon low angled ramps is not such a straightforward process, with the reverse flow having the profile of a group of waveforms (Pantin and Leeder 1987). The research presented herein explores the consequences of the incidence of density (and turbidity) currents upon obstructions to their forward motion. The main objectives of this thesis are to:

(i) generate a working model that describes the flows generated by the incidence of density currents upon low-angled ramps and other flow obstructions and,

(ii) compare this model with: (a) ancient turbidites that have unusual sequences of sedimentary structures and, (b) data obtained from recent turbidity current flows.

1.2   INTRODUCTION TO THE REVIEW

Turbidity currents and turbidites have been much studied since the pioneering work of Kuenen (1950) and Kuenen and Migliorini (1950). A wealth of data from theory, experiment and field observations exists, particularly concerned with the determination of the flow dynamics of broadly unidirectional, waning flows. By way of contrast, the study of density currents dates back to the early twentieth century, with workers considering the problems of the mechanics of cold squalls in the atmosphere (Schmidt 1911) and escaping gas in mines (Georgeson 1942). Many of the results from the theoretical, analytical and experimental study of density currents have been applied to the consideration of turbidity currents, thereby simplifying the system to one in which suspended load fall-out rate is not a variable. Exploration of the reversal of density currents by Pantin and Leeder (1987) resulted in the encompassing of internal solitary wave theory into the study of turbidity currents, thereby enlarging the scope for future study and forming the basis for this thesis.

The aims of this review are to examine the previous relevant work, highlighting areas in which there is a need for more study. Particular consideration will be given to:

(i) A review of the work upon natural turbidity currents with consideration of the occurrences of reflected turbidity currents and turbidites,

(ii) A review of the previous work on the mechanics of density and turbidity currents outlining the results that are important to this study,

(iii) An overview of the previous work and important results related to the study of bores (Simpson 1987) and cnoidal waves (Mysak 1984) with specific reference to the propagation of internal solitary waves (Mysak *op cit*).

1.3   THE PHYSICAL STUDY OF TURBIDITY CURRENTS AND TURBIDITES

1.3.1   The definition of turbidity currents

Much of the sediment that is removed from continental shelves and redeposited in the deep ocean is transported by sediment gravity flows rather than ocean currents, waves or tides.   "Sediment gravity flows" (Middleton and Hampton 1973) are mixtures of particles and water that move down slopes because the mixtures have a density greater than that of the ambient fluid.   Gravity acts on the solid particles in the mixture, inducing downslope flow.   The admixed water is a passive partner in this process.   The flow will continue to flow if the following conditions are satisfied: (i) the shear stress generated by the downslope gravity component acting on the excess density of the mixture exceeds frictional resistance to flow, and (ii) the grains are inhibited from settling by one of several support mechanisms (Table 1.1).   Four end member sediment gravity flows have been distinguished on the basis of the mechanism of grain suspension above the bed: (i) turbidity currents, (ii) grain flows, (iii) fluidized sediment flows and, (iv) debris flows (Fig. 1.1).   In reality, flows exist throughout a continuum between the end-members, with the end-members merely being a conceptual simplification.   Turbidity (density) currents are density flows in which the grain suspension is mainly supported by the upward component of fluid turbulence.

Bagnold (1962) observed that below a certain grain size, the sediment in a turbidity current flowing down an incline appeared to be supported by the turbulence.   This led him to suggest that under certain circumstances "the power provided by the tangential gravity component on the excess weight of entrained sediment is sufficient not only to maintain the suspension, but also to contribute towards the power needed to maintain the flow of the fluid drag exerted at the bed boundary".   Under such conditions the flow could maintain itself indefinately and was said to be in the hypothetical state of "autosuspension".   In such a state, a turbidity current will not erode or deposit sediment and a complete feedback loop exists such that:

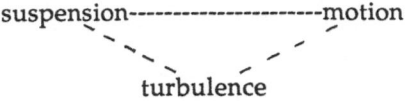

Middleton and Hampton (1976)

There is no documented evidence for the existence of autosuspension from laboratory work.   The Grand Banks turbidite of 1929 (Heezen and Ewing 1952) however, provides

Flow type	Major suport	Minor support
Turbidity current	1	3,4 (high-concentrations)
Liquefied flow	4	3,1
Debris flow[a]	5,2,6,4,3	1 (very large flows)

[a]Debris flow mechanisms from Pierson (1981)

Table 1.1

Support mechanisms for sediment gravity flows of significance in marine transport. 1 = turbulence, 2 = buoyancy, 3 = grain collisions (dispersive pressure), 4 = trapped or escaping pore fluids, 5 = frictional strength, and 6 = cohesive strength. From Pickering, Hiscott and Hein (1989).

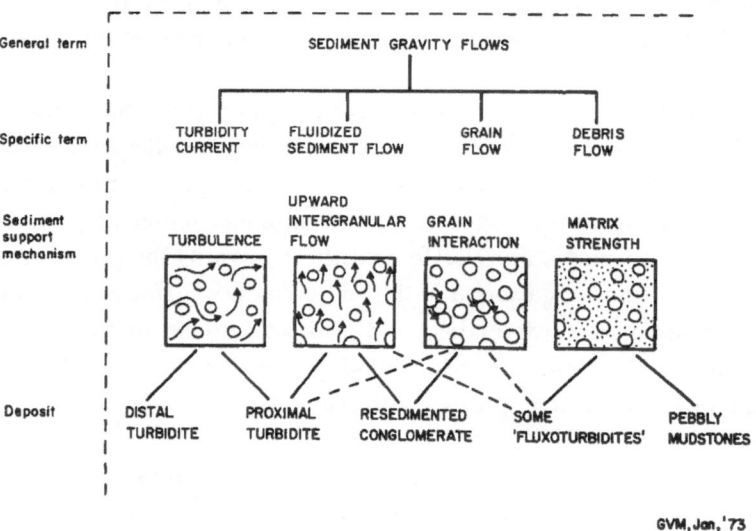

Fig. 1.1

Classification of subaqueous sediment gravity flows. From Middleton and Hampton (1976).

evidence of turbidity currents travelling a great distance (a minimum of 648 km.) providing potential evidence for autosuspension in natural turbidity currents.

Turbidity currents may be relatively short events (eg. the Grand Banks turbidity current of 1929 described by Heezen and Ewing 1952), in which a limited sediment source was generated, perhaps by an earthquake inducing slope failure. Such a current is called a *surge* type flow. Alternatively, turbidity currents can be quasi-continuous events due to the prolonged input of sediment from a long-lived source, such as a river mouth, and are entitled steady- or uniform underflows. (Fig. 1.2).

1.3.2 The vertical sequence of sedimentary structures and facies models associated with turbidites

The early experiments of Kuenen (1950) and Kuenen and Migliorini (1950) produced graded beds from deposition of high concentration mixtures of sand and mud. Kuenen (1950) discussed the possible form of such a deposit in nature placing emphasis upon recent oceanographic deposits. However, it was Kuenen and Migliorini (1950) who explained graded Oligocene greywacke beds in the Appenines in terms of deposition by turbidity currents. These two papers, along with Natland and Kuenen (1957), were the basis for the 'explosion' of study upon turbidity currents which effected the subsequent re-interpretation of many sequences that had previously been described as shallow water deposits (eg. Kuenen and Humbert 1964).

Bouma (1962), conducted a detailed sedimentological survey in the French and Swiss Alps and recognised that turbidite sequences consisted of a "deposition type" of five "intervals": (A) graded, (B) lower parallel laminae, (C) current ripple laminae, (D) upper parallel laminae and (E) pelitic. This sequence became known as the Bouma sequence (Fig. 1.3). Walker (1965) suggested that the term "division" replace the term "interval" used originally by Bouma (1962). Further study suggested that the complete Bouma sequence was rarely observed, but that it included all the lithologies and sedimentary structures in the order in which they tended to occur within the individual units (Dzulynski and Walton 1965), ie. it was a composite sequence. The "classic" Bouma sequence applies mainly to medium-grained sand grading to muds and silts, and according to the model of Walker (1978) is a result of deposition upon a fan lobe of a submarine fan (fig 1·4)Studies expanded to include depositional sequences that resulted from turbidity currents in which physical conditions did not allow the deposition of the ideal Bouma sequence. Such currents included high-density turbidity currents that contained a high percentage of cobbles to coarse-grained sand (Lowe 1982), and fine grained turbidity currents that were originally incorporated into the Bouma (E) division (Piper 1978).

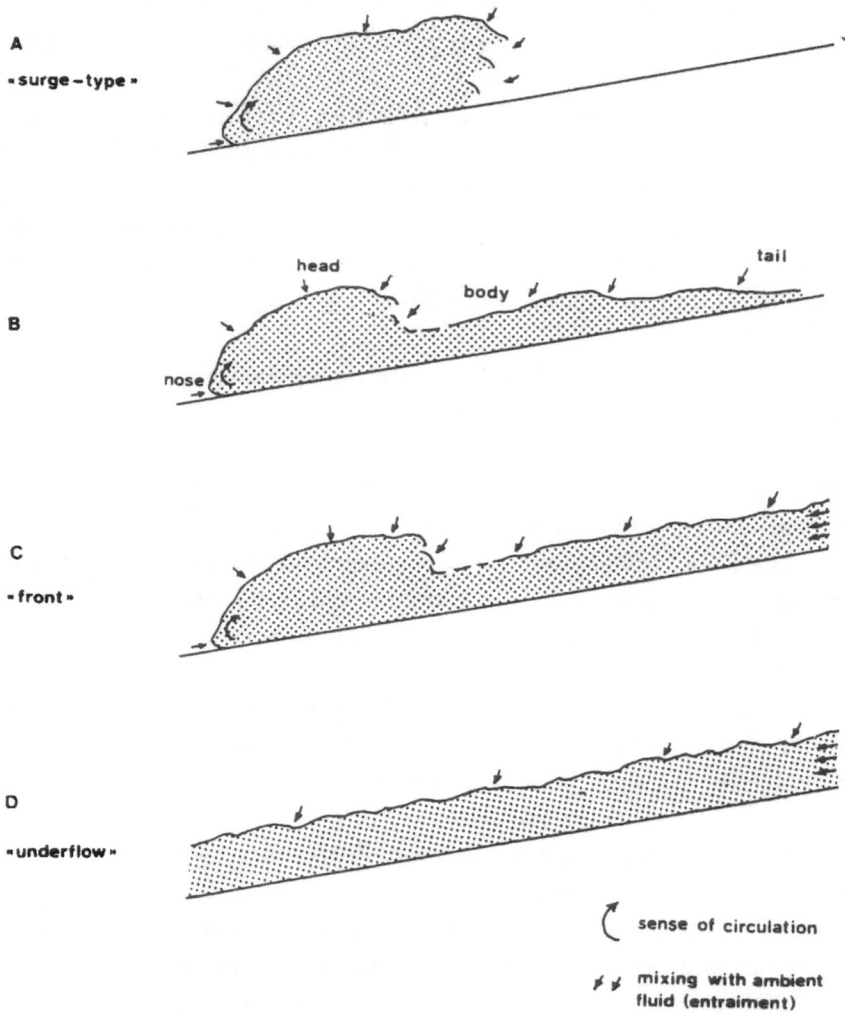

Fig. 1.2

Sketch of different types of turbidity currents. From Luthi (1980) (A and C after Hopfinger and Tochon-Danguy 1977).

GRAIN SIZE	BOUMA (1962) DIVISIONS	INTERPRETATION
Mud	E Laminated to homogeneous mud	Deposition from low–density tail of turbidity current ± settling of pelagic or hemipelagic particles
Silt	D Upper mud/silt laminae	Shear sorting of grains & flocs
Sand	C Ripples, climbing ripples, wavy or convolute laminae	Lower part of lower flow regime of Simons *et al* (1965)
Sand	B Plane laminae	Upper flow regime plane bed
Coarse Sand	A Structureless or graded sand to granule	Rapid deposition with no traction transport, possible quick (liquefied) bed

Fig. 1.3

Ideal sequence of sedimentary structures in a turbidite bed. From Pickering, Hiscott and Hein (1989) (after Bouma 1963, with interpretations after Harms and Fahnestock 1965, Walker 1965, Middleton 1967, Walton 1967, Stow and Bowen 1980).

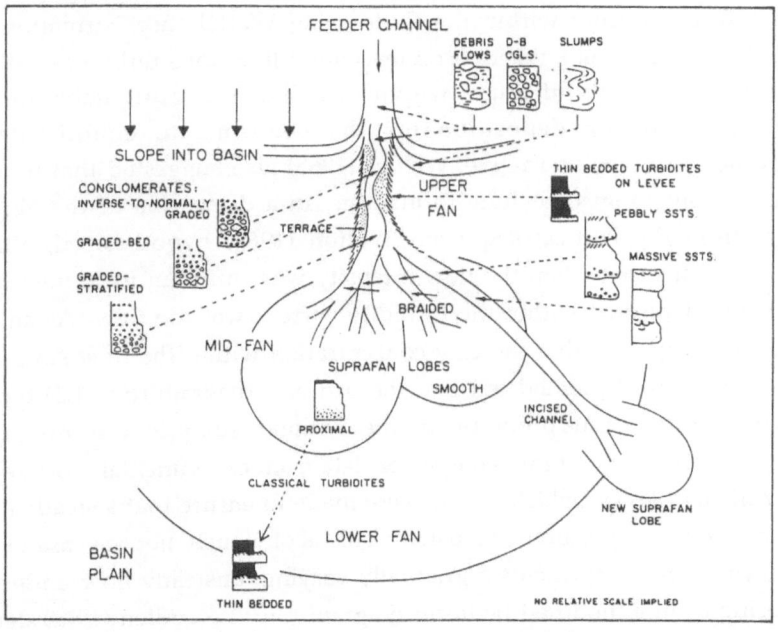

Fig. 1.4

Classic submarine fan model proposed by Walker (1978). Facies defined in ancient rocks are shown in their inferred position on the fan.

Early comparisons of ancient and modern turbidites were completed by Kuenen (1964), however, it was Walker (1978) who developed the first comprehensive facies model for deposition of turbidites on deep-sea fans (Fig. 1.4). The model included the "ideal" Bouma sequence, as well as sequences for coarse and fine grained turbidite deposits, and in so doing, developed the concept of proximal and distal facies associations.

1.3.3   Hydrodynamic interpretations of the Bouma sequence

One of the first hydrodynamic interpretations of the Bouma sequence was developed by Walker (1965) who used measurements and observations made in flume experiments, under conditions of non-deposition and varying flow velocity. Walker (*op cit*) divided the deposition of an ideal 'Bouma type' turbidite into seven hydrodynamic stages. These stages were related to deposition from a waning flow. Walton (1967) stated that the observed sequence of structures were related to the decrease in capacity and competency of the turbidity current flow. In addition, Walton (*op cit*) suggested that the development of a particular sedimentary structure was controlled by the grain size available, the rate of change of deceleration and the velocity at that point.

Dunes were not a part of the hydrodynamic sequence of Walker (1965) although their presence is apparent between the plane bed and ripple phases in the experiments of Simons *et al* (1961). One of the rare occurrences of dune scale cross-stratification in the sedimentary record occurs within the Ordovician Whitehouse Formation at Girvan, Scotland, U.K. (Hubert 1966), where it occurs within limestone turbidites. Walker (1965) suggested that the paucity of dunes within turbidites was attributable to there being insufficient time during deposition for the bedforms to equilibriate with the decelerating flow. As an alternative, Walker (1965) also suggested that the flow depth may be insufficient to sustain dune formation. In a discussion of the absence of the dune phase from the vertical sequence, Walton (1967) hypothesized, although was unable to determine, that when the flow intensity was sufficient for dune formation the grain size being deposited within the turbidity current was too fine. As an alternative, Walton (*op cit*) suggested that the clay concentration within the flow suppressed dune formation. Allen (1970) pointed out that the bedform diagram (Fig. 1.5) typically used for the hydrodynamic interpretation of the Bouma sequence was not immediately applicable. The bedform diagram consists of data from experimental work (eg. Guy *et al* 1966 and Williams 1967),in which efforts were made to ensure that a steady state existed and there was no net deposition or erosion. This is obviously not the case in a decaying turbidity current which represents a gradually varying, unsteady flow under conditions of net deposition. In a modified bedform diagram (Fig. 1.6), Allen (1970) demonstrated that it was possible for a waning current to pass from upper stage plane beds through to current ripples , without the generation of a dune phase. The modified diagram

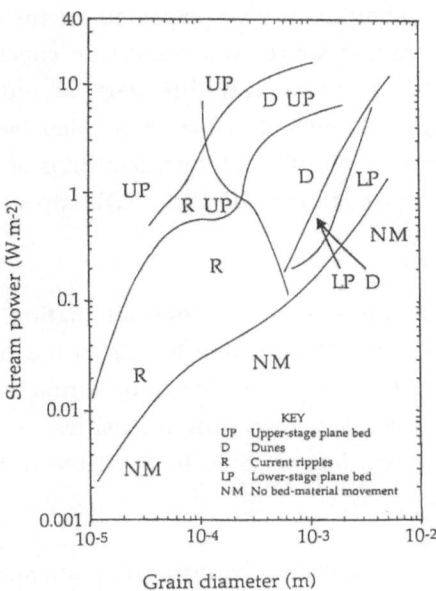

Fig. 1.5

Experimental existance fields for aqueous bedforms under equilibrium conditions shown in the stream power-grain size plane at 25°C. From Allen (1982).

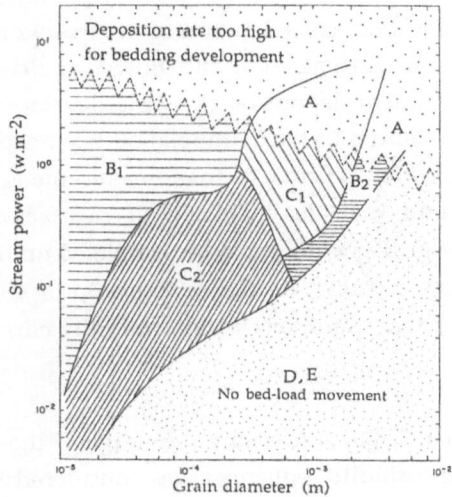

Fig. 1.6

Generalised stability-field diagram for current structures in turbidites proposed by Allen (1970). Modified from Allen (1982).

suggested that dune bedforms should only be expected in turbidity currents that originated in relatively coarse-grained source sediments. In conclusion, Allen (1970) suggested the following, modified, 'ideal' turbidite vertical sedimentary structure sequence: (A) graded division, (B1) lower division of parallel laminae (upper stage plane beds), (C1) cross bedding division, (B2) intermediate division of parallel laminae (lower stage plane beds), (C2) cross laminae division, (D) upper division of parallel laminae, and (E) mud division.

Another possible explanation for the scarcity of dune lamination is that as the flow intensity decreases, the high sediment fall-out rate results in the suppression of near-bed turbulence, which in turn inhibits ripple and/or dune formation and maintains an upper stage plane bed field at relatively low, flow intensities. Compounded with a sediment fallout 'lag' effect, these two factors could be the cause of the lack of dunes in turbidite sequences (Lowe 1988).

Palaeoflow velocities can be estimated from a bedform diagram such as that in Figure 1.5, by using the Bagnold criterion (Bagnold 1966) as verified by Allen and Leeder (1980) at the plane lamination (B) to ripple (C) division boundary. In addition, palaeoflow velocities can be determined from the settling velocity of grains within the flow by the application of the suspension criterion (Middleton 1976). Using a turbidite from the Capistrano Formation (Miocene to Pliocene) of California, Komar (1985) compared the two methods and found that the velocities obtained from the sedimentary structures were almost an order of magnitude greater than those derived using the settling velocities (Fig. 1.7). Both sets of results were reasonable and did not recommend one theory over the other. In addition, consideration of the many other factors that may be involved, including sediment concentration and flow turbulence effects on the grain settling velocities; lag of sediment deposition or ripple development in the decelerating flow; and the scale effects required in the extrapolation from experimental conditions to thick turbidity currents, provide no firm explanation as to the cause of the discrepancy. In fact, Komar (1985) found that the inclusion of the factors only worsened the discrepancy and stated that further work must be completed on the basic applicability of the suspension criterion and possibly, on the relationship of sedimentary bedforms to flow intensity before palaeoflow conditions within turbidity currents can be determined with any certainty.

The doubts raised by Komar (1985) concerning the usage of the bedform diagram (Fig. 1.5) in the interpretation of turbidite sequences was considered by Lowe (1988). Lowe (1988) pointed out that the assumption that any developing bedform depends only on the independent variables of flow intensity and grain size is acceptable for flows in which the bedload and suspended load are attained through tractive bed erosion (Lowe

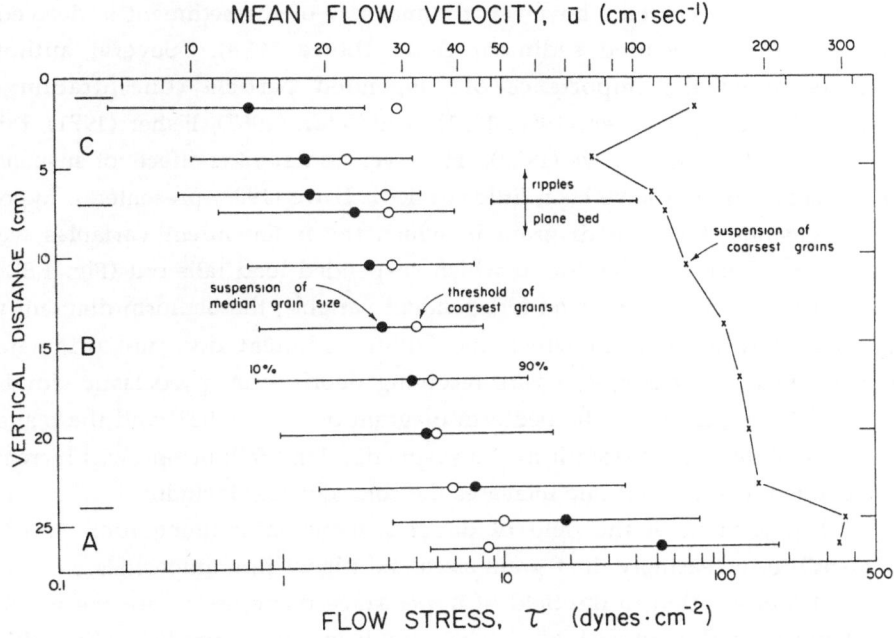

Fig. 1.7

Evaluations of the flow stress τ and mean velocity u at the time of deposition of a turbidite from the Capistrano Formation, California. The evaluations are based upon the suspension criterion, the threshold of the coarsest grains in the samples, and the transition from plane bed to ripples (Bouma B to C divisions). From Komar (1985).

1988). In turbidity currents however, the majority of the sediment is derived by the collapse of the suspended sediment cloud (Lowe 1988). Several authors have previously noted the importance of suspended particle concentration and its depositional rate, eg. Kuenen (1950, 1951), Middleton (1967), Fisher (1971), Middleton and Hampton (1976) and Lowe (1982). However, to date, the effects of suspensions on sediment concentration have been little studied. Lowe (1988) presented a hypothetical, three-dimensional bedform diagram in which the independent variables were flow intensity, grain size and the rate at which suspended load falls out (Fig. 1.8). As the suspended load fall-out rate is an independent variable, the bedform diagram will also be applicable to systems in which the falling sediment does not originate in the suspended load, for example, rivers receiving debris from pyroclastic clouds (Lowe 1988). The basal plane was the bedform diagram of Allen (1982) and the planes above show possible deviations from it as the suspended load fallout rate was increased. As the suspended load fall-out rate increases the consequences include:

(i) Expansion of the field of direct suspension sedimentation ('no bed-load movement') to increasingly finer grain sizes and higher flow intensities,

(ii) Modification to the field of upper stage plane beds with the generation of planar laminae at flow intensities considerably below those predicted by stability-field diagrams. This is a direct result of suppression of near bed turbulence by the fall-out of sediment,

(iii) Development of a Bouma D division in the finer grain sizes, separating the field of no bed-load movement from the ripple field and,

(iv) The dominance of upper stage plane beds under virtually all flow conditions for very fine-grained sands.

1.3.4 Hydraulic jumps in turbidity currents

Hydraulic jumps are associated with the change in flow regime from supercritical (Froude Number greater than 1) to subcritical (Froude Number less than 1) and *vice versa*. The quantitative study of the formation of hydraulic jumps in turbidity currents was initiated by Van Andel and Komar (1969). They suggested that in order to explain turbidite sequences observed in certain basins on the flanks of the Mid-Atlantic Ridge, a hydraulic jump was required to convert a thin, dense supercritical flow on the wall of a basin into a thick, less dense, low velocity, subcritical flow on its floor. Following on from this, Komar (1971) considered hydraulic jumps in turbidity currents that approached uniform flow conditions in sloping channels. He concluded that in a canyon of uniform gradient, a steady flow will be supercritical, whilst on the submarine fan the flow will be subcritical, with the change in flow regime occurring within the region near the canyon mouth. During the jump, the flow decreases in velocity, increases in thickness and increases in entrainment of sea-water (resulting in a decrease

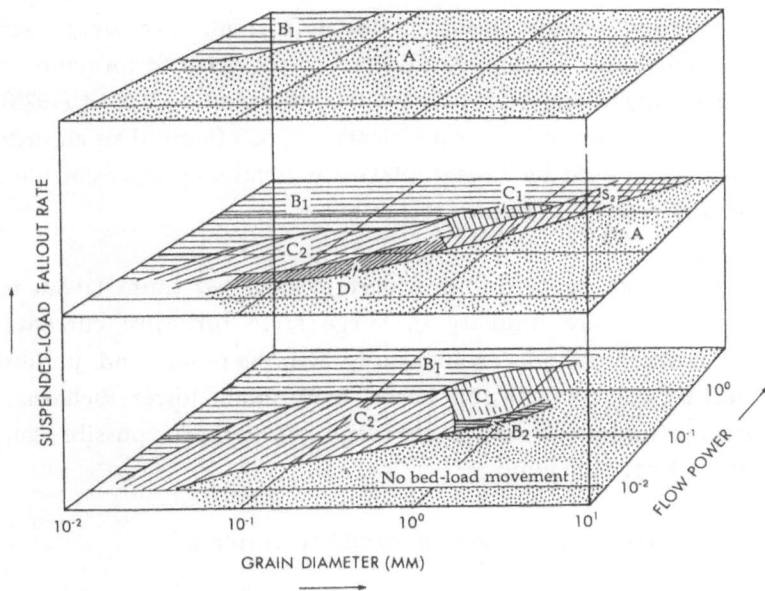

Fig. 1.8

Schematic, 3-dimensional stability-field diagram in which flow power, grain size and suspended-load fallout rate are treated as mutually independent variables. The generalised stability-field diagram for turbidites of Allen (1982) serves as the basal plane, corresponding to a condition of suspended-load fallout rate = zero. Two additional planes are shown representing intermediate (middle plane) and high (upper plane) values of suspended-load fallout rate. Modified after Lowe (1988).

in density) (Komar 1971). In contrast to this, Hand (1974) determined that hydraulic jumps and antidunes would form in a density current, of constant discharge, on slopes as little as 0.001°, and therefore would not be restricted to formation near the canyon mouth. These conclusions were subsequently disputed by Komar (1975) who suggested that Hand (1974) in using Drag coefficients of 0.005 (more than an order of magnitude greater than those used by Komar 1971) was obtaining unreasonable flow conditions that resulted in velocities of up to 150km h^{-1}.

An experimental study of turbidity current flow led Luthi (1980) to conclude that hydraulic jumps were unlikely in surge style turbidity currents in which the Richardson number (Ri) was around unity. On the other hand, underflows tend to be less diluted by entrainment and hence, may obtain lower Richardson numbers (ie. higher Froude numbers) than their surge counterparts possibly enabling them to accommodate hydraulic jumps.

1.3.5 Measurements from natural turbidity currents

Data upon the fluid dynamics of natural turbidity currents are sparse. The timing and location of turbidity currents are, as yet, unpredictable and upon occurrence the power of the flow often damages or destroys the equipment, making any study a hazardous and expensive affair. The best known turbidity current data was obtained from the cable breaks associated with the Grand Banks turbidity current of 1929 (Heezen and Ewing 1952). Since then several workers have tried to obtain data from turbidity currents with varying degrees of success and failure for example, eg. Inman *et al* (1976), Hay *et al* (1982), Bowen *et al* (1984), Piper *et al* (1985), Wright *et al* (1988) and Prior *et al* (1987).

1.4 THE EXPERIMENTAL MODELLING OF TURBIDITY CURRENTS

The limited success of the study of natural turbidity currents led several workers to try to experimentally model large scale flows, for example, Buffington (1961), Siengenthaler *et al* (1984). These studies met with limited success, giving information on the geometry of the flows, but not on the development of the physical nature of the flow or on the deposited sequence of structures. Flume studies that considered the deposition of sediment from suspension currents met with more success, and some of the results from this are outlined below.

1.4.1 Experimental investigations using suspended sediment

The majority of experimental research on density currents has been undertaken by hydraulic engineers using salt water solutions and fresh water. Much less research, has used density currents in which the density excess is provided by suspended sediment, although, notable exceptions include the work of Middleton (1967); Riddell (1969) and Luthi (1980). The problems in experimental investigation of sediment transport, deposition and erosion have been discussed by several authors for instance, Middleton (1966a), Simons *et al* (1965). The biggest problem encountered is in scaling down the experimental variables from the natural prototype. Middleton (1966a) suggested that the best way to do this was to ensure Froude number similarity between the experiment and the prototype, thereby keeping mixing and friction at the upper surface similar. Riddell (1969) generated surge-type, suspension, density currents and compared these with results from saline density current counterparts. These experiments demonstrated that suspension generated currents had a lower, more variable initial velocity than their saline solution equivalents and that for a particular density, the velocity of the suspension became greater as the proportion of fine-grained material was increased. Riddell (1969) observed that although fine grained suspension currents were most similar to saline solutions in their rate of velocity change and travel distance, coarse grained suspensions were more similar than fine-grained suspensions whilst the current was in its initial developmental stage.

Luthi (1980) released chalk suspension currents into fresh water in an experimental study that included the flow of surge style turbidity currents. He found that the flow thickness decreased with time; that the density distribution was uniform vertically through the current and that the Richardson number was approximately 1.5. This value of the Richardson number was slightly lower than that suggested by previous authors for example, 1.8 (Middleton 1967) and 2.0 (Kersey and Hsu 1976), who used the assumption that the density of the flow remained constant throughout the duration of the flow (Luthi 1980).

Parker *et al* (1987) conducted experiments to examine the behaviour of turbidity currents laden with non-cohesive silt (silica flour), moving down a slope, the bed of which was covered with similar silt. The currents were supercritical and maintained continuous flow conditions for the duration of the study. Although, only limited erosion and deposition occurred, ripples were generated over the surface of the bed. The erosive capabilities of a saline density current moving over a bed of loose, fine-grained sediment were studied by Garcia and Parker (1988). Results indicated that a strong density current could entrain considerable amounts of sediment, and that as well as the sand driven currents considered by Parker *et al* (1986), currents of silty mud could also involve major erosion and entrainment of sand, and carry it into deep water. Akiyama and Stefan (1988) considered numerically, an eroding and depositing current and expanded on the model of Parker *et al* (1987) by studying a laterally expanding turbidity current in which aggradation and degradation of the channel bed by sediment deposition and erosion by the expanding turbidity current were included.

1.4.2 Deposition of sediment

Experiments that used sediment suspensions in order to produce the excess density in turbidity currents were found to form graded beds (eg. Kuenen 1951 and Middleton 1967). Middleton (1967) used plastic beads in a series of experiments in which both distribution and coarse-tail grading were produced in the graded beds. However, the sediment was always deposited out of suspension far too quickly for ripples or ripple-drift lamination to form - both common features of natural turbidites. In experiments that distinguished between high and low concentration flows, Middleton (1967) observed that in the former, deposition occurred directly behind the head and resulted in the formation of a 'quick' bed which subsequently developed into a plane bed. In contrast to this, low concentration flows have a slow initial rate of sedimentation with little deposition behind the head. Regardless of the initial differences in sedimentation patterns, the final stage of sedimentation in both low and high concentration flows is a slow settling out from suspension (Middleton 1967).

Applying different rates of deceleration to suspension currents in a circular flume tank, Banjeree (1977) produced Bouma like sequences. At slow deceleration rates, Bouma-like sequences with the development of divisions A, B, C and E were generated and that upon increasing the rate of deceleration, suppression of the Bouma C division occurred. In addition, Banjeree (1977) determined that a graded bed was generated by an abrupt, rather than continuous deceleration and that if an initial, instantaneous deceleration, was not applied the first structure formed was either antidune- or parallel laminae.

Deposition of heterogeneous sediment from a suspension current that had been allowed to attain equilibrium was examined by Ghosh and Sengupta (1986). Their experiments demonstrated that upon deceleration, all grain sizes were simultaneously deposited - a result which they felt may go some way to explaining the presence of large proportions of fine-grained matrix in greywackes together with coarse framework grains. The lack of grading that is achieved in these experiments, initially contradicts the earlier experimental findings of Kuenen and Migliorini (1950). However, Kuenen and Migliorini (1950) ascribe the vertical grading to "dilute" suspensions that lag behind the main body of the turbidity current, and flow with a continuously decreasing velocity generating the observed vertical decrease in the maximum grain size. The experiments conducted by Ghosh and Sengupta (1986), in a re-circulating flume, do not fit the flow description of Kuenen and Migliorini (1950) and therefore, grading would not be expected.

The geometry of sand deposited by turbidity currents was examined by Middleton and Neal (1989), who produced experimental turbidites by releasing suspensions of spherical beads from a lock into a 6m long channel. The generated beds had a uniform thickness, except in the very distal parts where there was a rapid decrease in thickness. For equal volumes of sediment increasing the concentration of the flow decreased the distance of travel, but increased the thickness of the deposited bed. At lower concentrations grading was better developed and in high concentration flows, reverse grading may occur near the base of the bed.

1.5   THE MECHANICS OF DENSITY CURRENT FLOW

1.5.1   Introduction

A wealth of theoretical and experimental data exists in hydraulic and engineering research that is concerned with the motion of density currents. The predominant aim of this work has been to model a variety of engineering situations in which density current flow plays a fundamental role, for example, at the heads of reservoirs and near effluent sites into lakes (Johnson *et al* 1987; Akiyama and Stefan 1987). Some of the earliest work on gravity currents was carried out by Schmidt (1911) who used a laboratory water channel in order to model the front of an approaching cold squall in the atmosphere. Renewed studies in the 1950's led to an understanding of many of the properties of turbulent density flow, the results of which can be applied to considerations of turbidity current flow and are outlined below.

Density currents can be generated by the sudden release of a finite amount of denser fluid into one that is less dense (and *vice versa*). After rapidly attaining a maximum velocity, such currents will usually be slowed down by friction and the entrainment of the less (or more) dense ambient fluid. Regardless of whether the density current is a surge or steady type of flow, they begin life as a surge which rapidly develops into current with a head, body and tail (Fig 1.9). Density currents have been modelled in the laboratory by the sudden release of a denser fluid (eg. salt-water, clay suspensions or sediment-water mixture) into a channel that is filled with a less dense fluid (usually cold, fresh water).

1.5.2   The motion of the head of a density current

1.5.2.i   Theoretical considerations

Early theoretical studies of the hydraulics of density currents were completed by Von Karman (1940). Von Karman (1940) considered, theoretically, the flow of a non-

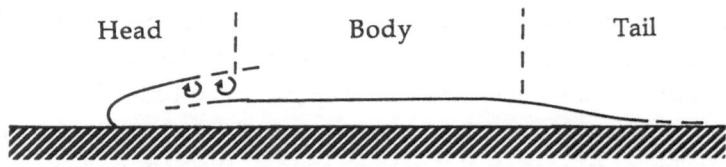

Fig. 1.9
Schematic division of a turbidity current. Modified after Middleton and Hampton (1976).

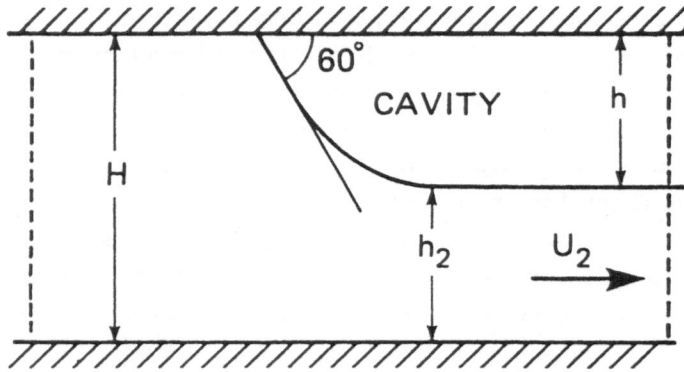

Fig. 1.10
The theoretical treatment of a gravity current head used by Benjamin (1968). The fluid is displaced as it passes under the cavity. From Simpson (1988)

dissipative denser current into a less dense fluid at great depth. Using Bernoulli's Theorem in a form applicable to steady inviscid flows he deduced that the interface would make an angle of 60° at its base.

In order to reduce the problem of the motion of a gravity current to its simplest form, Benjamin (1968) analysed the flow past an empty or air filled cavity that displaced liquid beneath a horizontal boundary (Fig. 1.10). To simplify the analysis, viscosity and surface tension were ignored, ie. an inviscid fluid system was considered, and a condition of no energy loss from the system was assumed. Application of Bernoulli's Theorum along the free surface produced the following relationship:

$$u^2 = g(H^2 - d'^2)H / (2H - d')d' \qquad (1.1)$$

where u is the velocity past the cavity a long way downstream of the stagnation point, O; g is the acceleration due to gravity; H is the total depth of the fluid and d' is the fluid depth beneath the maximum height of the cavity. The only non-trivial solution, that satisfied the assumption of no-energy loss, was when d' was equal to 0.5H. The effect of a free-surface (the theoretical removal of the lower surface in the cavity flow), was to reduce the height of the current (d) such that $d/H = 0.5 (1 - 15/16\xi)$ where $\xi = \Delta\rho/\rho$ ($\Delta\rho$ is the positive density difference between the two fluids and ρ is the density of the fluid approaching from ahead). Allowing for uniform loss of energy from the head resulted in the following relation:

$$\Delta = (2d' - H)(H - d'^2 / 2d'(2H - d') \qquad (1.2)$$

where Δ represents the uniform loss of energy from the head. The relationship showed that when d'/H < 0.5 a steady flow could not be achieved without the input of additional energy (Benjamin 1968). Further consideration of fractional depth (d'/H) led Benjamin (1968) to suggest that, within the range of 0.5 < d/H < 0.6527, the flow of the ambient fluid would be supercritical and a steady flow unlikely to exist. In conclusion, the head of a density current was a vital feature of the dissipative process for any situation in which d was not equal to 0.5H. Extending the analysis to situations in which d/H approached zero (ie. applicable to the study of turbidity currents in the ocean) Benjamin (1968) demonstrated theoretically, that vigorous breaking must occur on the rearward side of the head in order to dissipate the excess energy. This conclusion agreed well with experimental observations on gravity currents. In addition, Benjamin (1968) determined that the propagation velocity of the head of the density current (U_h) was:

$$U_h = [2gh_2(\Delta\rho/\rho)]^{1/2} \qquad (1.3)$$

where h_2 is the mean thickness of the density flow behind the head.

Later authors point out the limitations of Benjamin's (1968) analysis, as applied to real situations. For instance, the effects of mixing between the two fluids were too important to simply be ignored (Britter and Simpson 1978). Also, the use of an inviscid analysis ignored non-negligable effects that occurred as a result of friction at the rigid boundary, eg. the over-running of some of the ambient fluid at the base of the gravity current (Allen 1971; Simpson 1972).

1.5.2.ii Experimental investigations

Keulegan (1957, 1958) studied surges of saline solutions from locks into horizontal channels filled with fresh water. From measurements of the thickness of the head (h_1), velocity of advance (U_h) and the density (ρ_1) of the head (Fig. 1.11) he established the 'specific law of saline heads' which, when the Reynolds number of the flow was greater than 1200, could be represented thus:

$$U_h = 0.71 \left[\frac{((\Delta\rho)gh_1)}{\rho_m} \right]^{1/2} \qquad (1.4)$$

where $\Delta\rho$ is the difference in density between the head and the ambient fluid ($\Delta\rho = \rho_1 - \rho_2$); ρ_m is the average values of ρ_1 and ρ_2, and g the acceleration due to gravity. The constant of proportionality was found to be dependent upon the Reynolds number of the flow. However, Keulegan (1958) showed that, providing the Reynolds number was greater than 1200, the variation in the "constant" was negligible. In an experimental re-examination of Keulegan's theory, Kersey and Hsu (1976) found that the constant (0.71) in equation 1.4 was not fixed but varied with the magnitude of the frictional losses; the steepness of the slope down which the current flowed and the fractional depth (h_1/H). Nevertheless, for flow over a horizontal surface Kersey and Hsu (1976) obtained the same value as Keulegan (1957, 1958) of 0.71, both experimentally and also theoretically.

In a detailed experimental examination of the motion of the head of a gravity current over a horizontal surface, Simpson and Britter (1979) discovered that with increasing fractional depth (h_2/H), Fr and U_h increased (Fig. 1.12); that in high Reynolds number flows the nose height was approximately constant at 1/8 of the total height of the head and that the mean velocity in the flow behind the head was approximately 1/6 greater than the velocity of advance of the head.

The initial stages of development of a gravity current formed by the sudden release of a

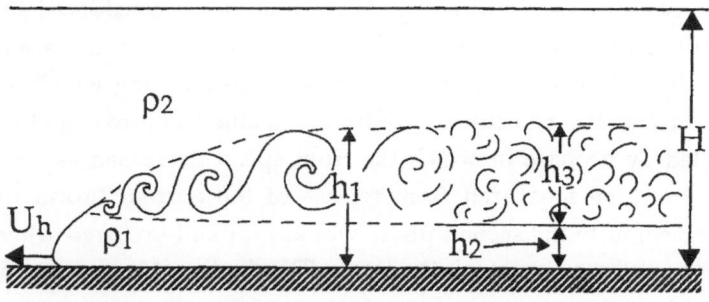

Fig. 1.11

General features and nomenclature of underflowing density currents.
Modified after Allen (1982).

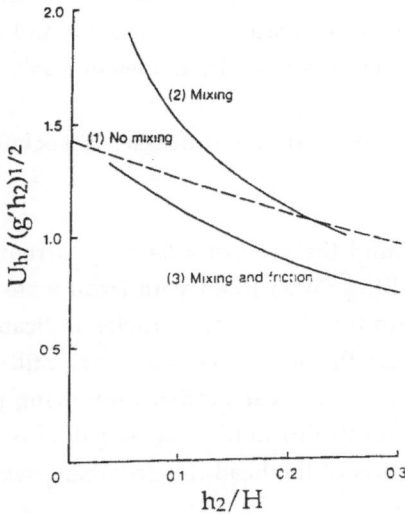

Fig. 1.12

The speed of the head of a gravity current advancing into fluid of
different depths. The speed is expressed as the Froude number and the
depth as the fraction of the current of the total depth. Curve (1) is the
inviscid current with no mixing, (2) is an inviscid current with mixing
and (3) is a mixing current moving along the ground. From Simpson
(1988). Modified after Simpson and Britter (1979).

finite volume of salt-water into a less dense fluid were considered by Huppert and Simpson (1980). Experiments demonstrated that, if viscous effects were negligable, then the current passed through an initial adjustment phase, during which initial conditions were important and the velocity of the front remained approximately constant. This was superseded by a phase in which the front speed decreased as a function of $t^{-1/3}$ where t was the time measured from release of the current (Rottman and Simpson 1983). The transition to the second phase was abrupt and occurred when a disturbance generated at the end-wall overtook the front. The depth of denser fluid in the lock (h_0) could be varied, as a fraction of the total depth of fluid (H), between 0 and 1. When h_0/H was equal to or slightly less than 1, the disturbance had the appearance of a hydraulic jump (Fig. 1.13). For smaller values of h_0/H the disturbance was a long wave of depression (Fig. 1.14) (Rottman and Simpson 1983). For the particular case of $h_0/H = 1$, the front was overtaken by the bore when it was approximately 10 lock lengths from the end-wall of the lock (Simpson 1982).

Laval *et al* (1988) also experimentally explored surge style density currents. Results indicated that the surge velocity was proportional to the square root of the initial height of the flow; to the square root of the ratio of density difference to the ambient fluid density, and increased also with an increase of the gradient on low slopes (1° to 7°). The first two factors had previously been verified by Keulegan (1957, 1958).

1.5.3 The internal structure and mixing processes associated with the head of a density current

To examine motion in and around the head of a density current, Middleton (1966b) ran a series of experiments in a tilting flume filled with fresh water, into which a constant supply of saline fluid was introduced. Tracer particles indicated a strongly divergent flow within the head. Behind the head, the flow was still found to be turbulent however, it was more uniform, with tracer particles following paths almost parallel to the bed (Fig. 1.15). Middleton (1966b) noted that the flow over a horizontal bottom resulted in the gradual dissipation of the head due to mixing with the ambient fluid.

The head of a gravity current is characterised by the occurrence of lobes, that are separated by clefts, which extend back into the flow as tunnels (Fig. 1.16). Lobes and clefts were examined experimentally by Simpson (1969, 1972) who found that they form a perpetually changing system in which the clefts extend back into the head a considerable distance. Clefts are separated from each other by lobes, the sub-division of which generates new clefts. The formation of lobes and clefts is a consequence of the no-slip condition that exists at the bottom boundary (Allen 1971) and as a result incurs the over-running of a thin layer of less dense ambient fluid. This gravitationally

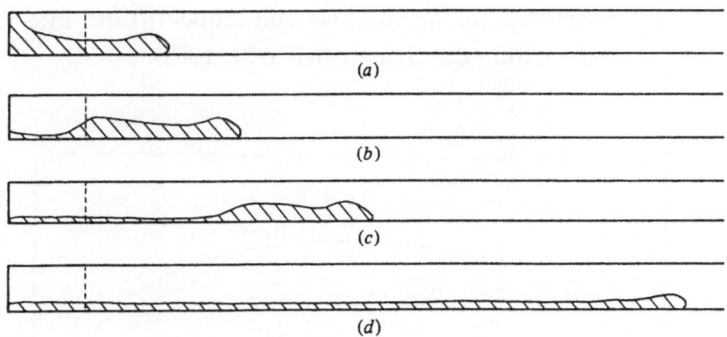

Fig. 1.13

Schematic illustrations of the collapse of a volume of heavy fluid with $h_0/H = 1$ at four successive times after release. From Rottman and Simpson (1983).

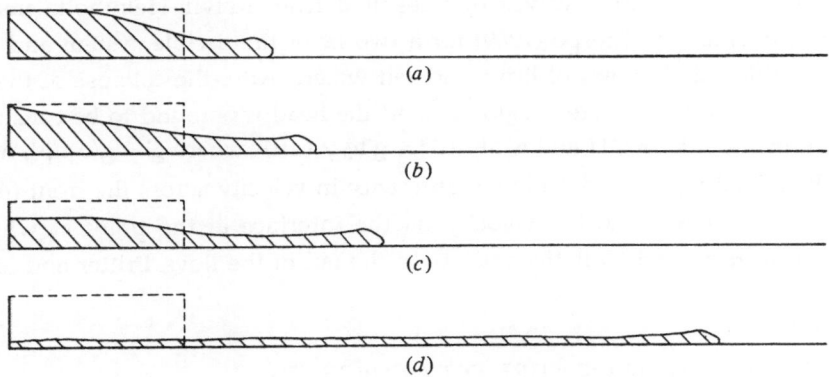

Fig. 1.14

Schematic illustrations of the collapse of a volume of heavy fluid with $h_0/H \ll 1$ at four successive times after release. From Rottman and Simpson (1983).

unstable system then inverts by mixing the over-run ambient fluid into the head, via corkscrew vortices that exist within the clefts (Allen 1971, 1985).

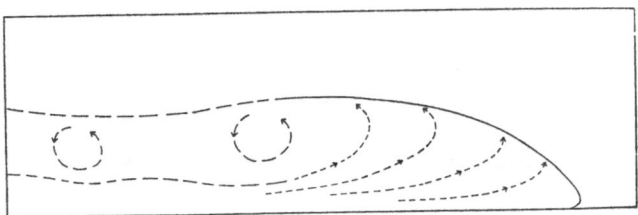

Fig. 1.15
Internal motion within the head of a density current. From Middleton (1966b).

Britter and Simpson (1978) examined the mixing at the head of a density current both experimentally and theoretically. For a flow in which there was no friction at the floor, ie. no over-running of ambient fluid by the gravity current, it was observed that billows began to form as small waves part way up the interface of the head (Fig. 1.17). The waves grew in amplitude until they formed distinct billows that were then convected away from the current. In any one experiment the billows grew to a particular size before they subsequently broke down to form a region of fine scale turbulence (Britter and Simpson 1978). The billows were thought to be a result of finite amplitude Kelvin-Helmholtz instability that results from the velocity discontinuity across the interface. The formation of the internal waves that resulted from Kelvin-Helmholtz instability was well documented by Thorpe (1969) for a two layer immiscible system and Thorpe (1971) for stratified shear flows of brine and salt water. After the collapse of the billow structure, the height of the mixed region behind the head was found to be constant and independent of the ratio h_2/H and g' at $0.33 = g'h_3 / (\Delta u)^2$ where $g' = g(\Delta \rho / \rho)$, h_3 is the height of the mixed region and Δu is the difference in velocity across the front ($\Delta u = u_m - v_4$, where U_t, where u_m is the velocity on the interface at the point at which the billows begin to form and U_t is the velocity in the tail of the flow, Britter and Simpson 1978).

1.6   THE REFLECTION OF TURBIDITY CURRENTS

1.6.1   Some examples of the reversal of turbidity currents

The majority of experimental, theoretical and field studies of turbidity currents and turbidites have concentrated upon considerations of an essentially, unidirectional flow.

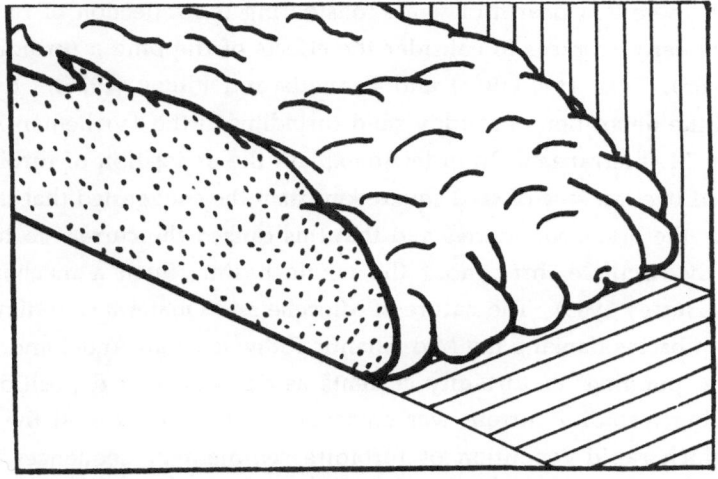

Fig. 1.16
Schematic representation of lobes and clefts at the front of a density current head moving along a horizontal surface. From Simpson (1988).

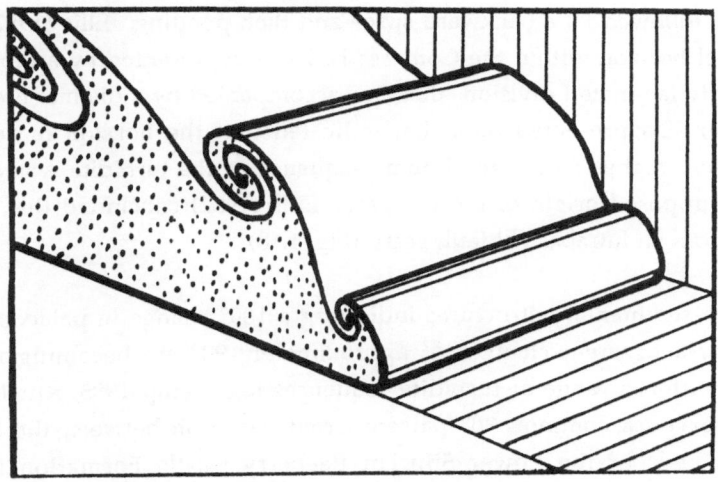

Fig. 1.17
Schematic representation of billows forming at the front of a density current head moving along a horizontal surface. From Simpson (1988).

As a result, there is a dearth of work considering the reflection of turbidity currents. Some of the early workers to consider the effects of trapping a turbidity current were Hersey (1965), Ryan *et al* (1965) and Bennetts and Pilkey (1976). Ryan *et al* (1965) considered the deposition of muddy sand turbidites in the Tyrrhenian Abyssal Plain of the western Mediterranean. In order to explain the deposition of turbidites that could be correlated over an area of 4662 square km basin they suggested that the currents were dammed by the basin boundaries and that this caused the current to turn back and to subsequently circulate throughout the whole basin. Such a mechanism is termed "ponding" (Hersey 1965). The nature of Miocene and Quaternary sedimentary turbidite sequences in basins flanking the Mid-Atlantic Ridge led Van Andel and Komar (1969) to also invoke "ponding" of turbidity currents as an important depositional mechanism. A rebounding turbidity current was envisaged to have deposited the majority of the sand in which rapid repetition of turbidite sedimentary sequences were observed. Subsequent to that, a reflecting internal seiche or bore could have generated the structures and the sorting observed in the calcareous oozes (Van Andel and Komar 1969).

The Contessa bed is a mega-turbidite unit that occurs within the Miocene Marnoso-Arenacea Formation of the Northern Appenines and was deposited in a NW-SE fault bound basin (Ricci-Lucci and Valmori 1980). Unlike the sequence of thinner basin plain and fan-fringe deposits with palaeocurrents from the NW within which it occurs, the Contessa has palaeocurrents from the SE (Ellis 1982). Parea and Ricci Lucci (1975) observed current ripples within the bed that suggested a 180° flow reversal from the flutes on the base of the bed. To explain this they suggested up fan climbing of the mega-flow followed by a backward surge and then ponding. Ellis (1982) suggested that the reversal horizon within the Contessa bed was represented by a blocky, 10 to 15 cm thick, poorly laminated division that was accompanied by a distinct break in grain size. Ellis (1982) also presented data that indicated that the horizon of reversal occurred progressively higher within the Bouma sequence as the turbidite was observed further from the supposed origin of the reversal. Ellis (1982) concluded that the current was reflected from an intrabasinal fault scarp (Fig. 1.18).

Apart from sedimentary structures indicating a 180° change in palaeocurrent direction, intrabed palaeocurrent changes of approximately 90° are becoming more commonly observed within a range of turbidite sequences (eg. Kemp 1985, Kneller 1991). Kemp (1985) observed a common 90° palaeocurrent variation between the basal flutes and ripples in beds of the Lower Silurian Raeberry Castle Formation in the Southern Uplands, Scotland. Deposition of these sediments was suggested to have occurred within a trench and it was suggested that the intrabed variation could be the result of the interaction of axial and trench slope currents. Similar palaeocurrent deflections are

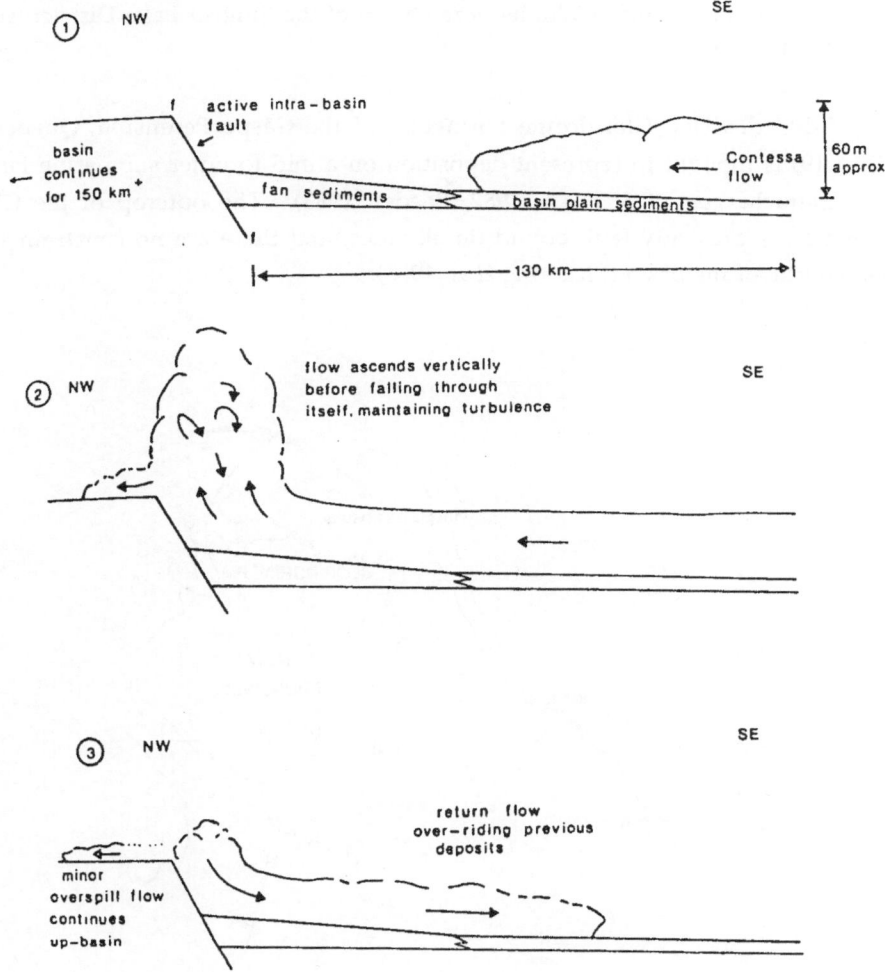

Fig. 1.18

Postulated flow reversal model for the Contessa turbidity current. From Ellis (1982).

observed in the Ordovician Windermere Group of the English Lake District (see chapter 3 and Kneller 1991).

The Mid-Ordovician Cloridorme Formation of the Gaspe Peninsula, Quebec, Canada (Fig. 1.19) is thought to represent deposition on a mid to outer submarine fan (Skipper and Bhattachargee 1978, Hesse 1982, Beeden 1983). The outcrop of the Cloridorme Formation is presently fault-bound on all sides, and there are no constraints as to the original size of the basin (Pickering *et al* 1986).

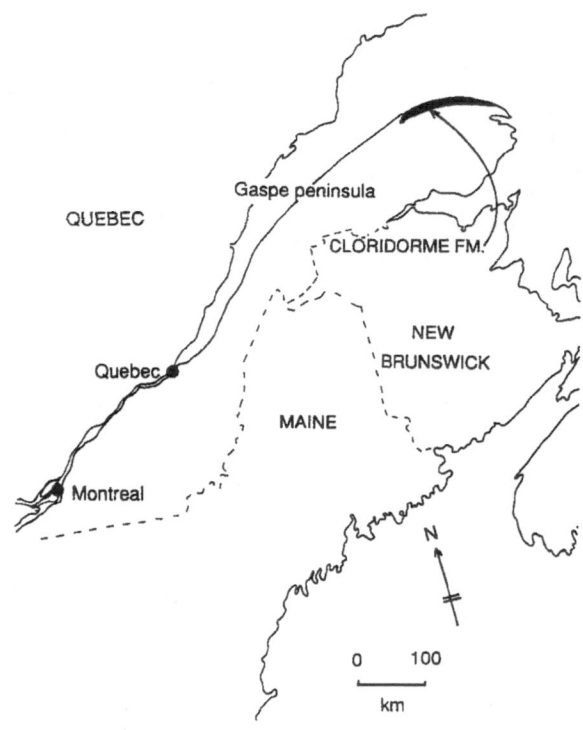

Fig. 1.19
Location map for the Cloridorme Formation, Gaspe Peninsula, Quebec, Canada.

The sedimentology has been described in detail and the Cloridorme Formation divided into seven members : β_1 to β_7 (Enos 1969). The lowest members (β_1 and β_2) have many thick calcareous bi-partite turbidites in which a lower coarse- to fine-grained sandstone fines up into a thick silty mudstone cap. Individual beds often contain palaeocurrent indicators suggesting both the usual palaeoflow direction of east to west and in addition, west to east flow. These beds have been interpreted as the result of deposition from antidunes (Skipper 1971, Hand *et al* 1972) and more recently as the result of the reflection of the original turbidity current flow (Pickering and Hiscott 1985; Pantin and Leeder 1987).

1.6.2 Experimental studies of the reflection of density currents

The reflection of experimental density currents from the vertical end-walls of flume tanks has often been noted (Allen 1985, Simpson 1987), but little qualitative and quantitative work on the phenomenon exists.

Muck and Underwood (1990) considered the run-up of density currents on a slope prior to the return of the reflected current. Their numerical and experimental studies suggested that the most important variable affecting the upslope flow of a turbidity current was the flow thickness. It was suggested that the maximum run-up elevation of a sub-critical turbidity current, would be approximately equal to 1.53 times the flow thickness.

Pantin and Leeder (1987) completed several exploratory runs which involved the reflection of a saline solution density current from a ramp in a rectangular flume tank filled with fresh water. Upon release, the surge developed the characteristic head, body and tail of a density current before being reflected from a ramp that could be varied between 7° and 10°. Upon reflection, a bulge of fluid was generated that moved in the reverse direction and broke down into a series of smooth, symmetrical waveforms. Analytical examination of these waveforms led Pantin and Leeder (1987) to deduce that the reversal of density currents resulted in the generation of a train of internal solitary waves. Such an observation contrasted with the assumption that the reversal of a density current and/or turbidity current generated a reverse flow that was identical in form to the original forward flow.

1.7 THE HYDRODYNAMIC STUDIES OF BORE FORMATION

A phenomenon related to the density current is the bore which, like density currents, are important for mass transfer. The best known type of bore is a tidal disturbance which moves upstream in some rivers. It is an example of a hydraulic jump in which there is a sudden increase in water depth associated with a change in flow regime from super- to subcritical, or *vice versa* (Simpson 1987). In addition to these moving jumps, stationary ones also exist.

Energy is lost at the jump either by entrainment of air, as in a turbulent bore, or by waves which carry the energy away as they move away from the front (Benjamin and Lighthill 1954). The wave train associated with a weak bore was examined theoretically by Benjamin and Lighthill (*op cit*). It was found that the wave train was cnoidal to a good approximation and that, for stronger bores, in the range in which breaking did not occur, the wave train closely resembled a series of solitary waves (Benjamin and Lighthill *op cit*). Wood and Simpson (1984) demonstrated that when the depth change across the bore was small, both the shear stress between the layers and the interfacial slopes were sufficiently small for the entrainment to be negligable. In this instance, the jump to resembled an undular jump in an open channel. Alternatively, for large depth changes, a shear flow instability occurred and there was much breaking associated with the jump (Wood and Simpson *op cit*).

A similar hydraulic effect may occur within stratified fluids and is known as an internal bore. The features of the internal bore are identical to the surface bore, the only difference being that the reduced gravity acting on the system causes a slower motion (Simpson 1987). Internal bores are a common natural phenomena being abundant in the atmosphere (eg. Smith 1988). They are also caused by tidal effects on fresh water layers near the coast (eg. Thorpe *et al* 1983); sea breeze fronts (Simpson 1987) and by the flow of air over mountains (Baines 1987). As a result, internal bores have been the subject of much experimental and theoretical study, some of which is outlined below.

A bore generated by the flow of a fluid over topography, or by the flow of a basal density current, was most effectively modelled by the examination of the flow of two fluids over a towed obstacle (eg. Long 1954, 1970; Houghton and Kasahara 1968; Baines 1984; Wood and Simpson 1984; Baines and Davies 1980). Long (1954) was able to determine, experimentally and analytically, regions in which a range of flow phenomena, including hydraulic jumps, occurred. Further studies examined these regions more closely and four categories of flow were subsequently identified, based upon the velocity of the flow and the height of the obstacle (Fig. 1.20). The boundaries of this diagram had been partially verified by the earlier experiments and theoretical considerations of Long (1970); Houghton and Kasahara (1968) and Baines (1984). Where the height of the obstacle approached or exceeded the basal layer, a hydraulic jump was generated upstream. Partial blocking may exist in which the flow above the obstacle was reduced and some of the fluid was forced to move ahead of the obstacle, ie. blocking means the movement of a region of fluid upstream at the speed of the obstacle (Simpson 1987).

Wood and Simpson (1984) and also Rottman and Simpson (1989) considered the generation of a bore, by the towing of an obstacle along the base of a two-layered miscible fluid system. In hydraulic theory, internal bores are idealised as travelling abrupt jumps in the level of the interface between immiscible fluids (eg. Yih and Guha 1955). In practice however, quantitative hydraulic theory does not provide an accurate picture of the structure of a bore as a function of its strength, particularly in miscible

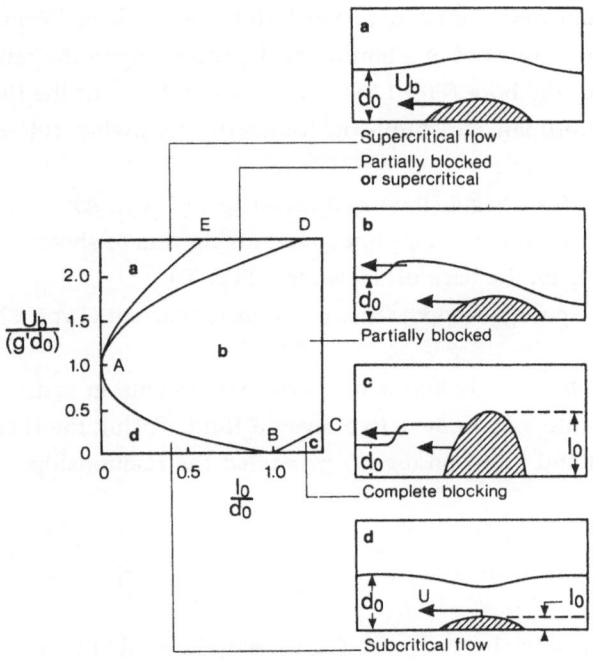

Fig. 1.20

The flow regimes for two-layer flow over a streamlined obstacle (with the upper layer of infinite depth). From Rottman and Simpson (1989), (adapted from Baines and Davies (1980).

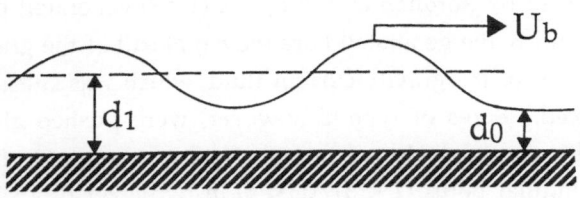

Fig. 1.21

Nomenclature for a jump moving into a stationary two-layer fluid.

fluid systems. Rottman and Simpson (1989) (after the work of Wood and Simpson 1984) found that the characteristics of a bore were dependent upon the ratio d_1/d_0, where d_1 is the mean depth of the bore (Fig. 1.21) and d_0 is the depth of the fluid prior to the arrival of the bore. Rottman and Simpson found the following rule-of-thumb to be applicable (Fig. 1.22a, b and c):

$1 < d_1/d_0 < 2$ -Bore has a smooth, undular form (Type A),

$2 < d_1/d_0 < 4$ -Bore is undular but some mixing due to shear instability occurs on the back of the waves (Type B),

$d_1/d_0 > 4$ -Bore has appearance of a density current (Type C)

Wood and Simpson (1984) conducted a series of experiments in order to examine the propagation of a bore into a stationary two-layered fluid. Using the theory of Chu and Baddour (1977), Wood and Simpson (*op cit*) generated the relationship:

$$\frac{U_b^2}{gd_1} = \frac{d_0' + d_1'(1-d_1)^2}{2d_0' - 3d_0'd_1' + d_1'^2} \tag{1.5}$$

where U_b is the velocity of the bore, $d_0' = d_0/H$; $d_1' = d_1/H$; d_0 depth of flow downstream of the the bore; d_1 is the depth upstream of the bore and H is the total depth (Fig. 1.21). Relation (1.5) can be plotted as in Figure 1.23 upon which the experimental data of Wood and Simpson (1984) is presented and also, the density current data from Simpson and Britter (1979). The latter is presented because when shear instability occurs, the bore behaves as a gravity current and as a result, the theory of Chu and Baddour (*op cit*) is no longer applicable.

Gravity currents, as they flow along a boundary or density interface, can generate internal bores upstream of themselves as shown by Maxworthy (1980) and Simpson (1982). The nature of the disturbance is dependent upon the relative densities and depths of the current and layer into which it is intruding. The flow has many similarities with the flow of a layered fluid over an obstacle, although it was not possible to generate a completely blocked flow (Rottman and Simpson 1989). Experiments conducted by Rottman and Simpson (1989) generated bore types A, B and C. In both type A and B, the generated bore moved ahead of the gravity current, taking with it a small amount of the gravity current fluid, which was subsequently left behind as the bore advanced. Bores of type C however, were pushed ahead of the density current and had the same velocity as the generating current. Rottman and Simpson (1989) found that undular bores ($1 < d_1/d_0 < 4$) in two layer fluids appear like gravity currents when $d_1/d_0 > 4$ but behave like them when $d_1/d_0 > 2$. Theoretically considering the motion of a density current beneath a two-layered system, Holyer and Huppert (1980) determined that if the inflow into a density current was sufficiently low,

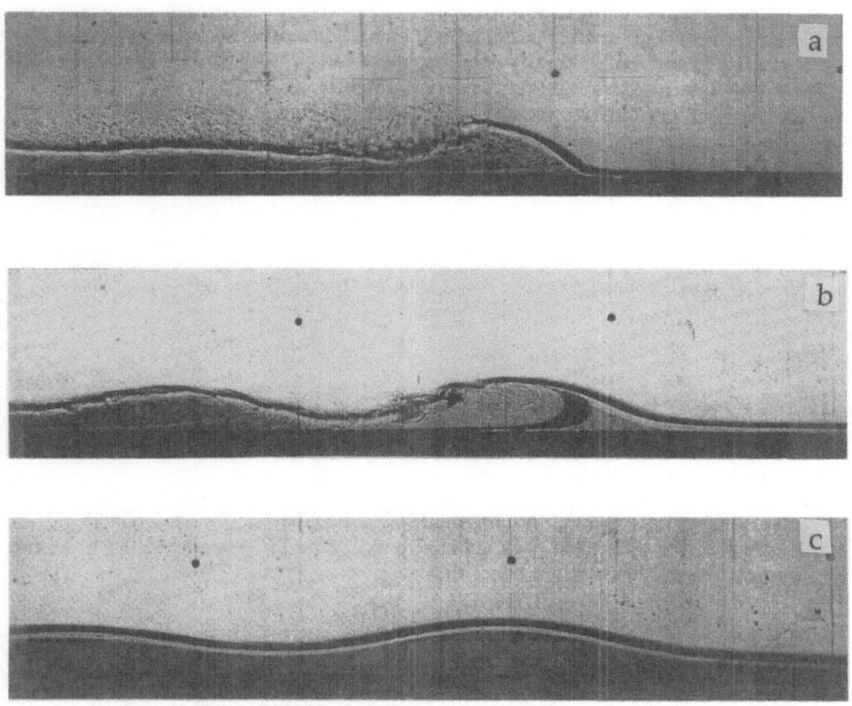

Fig. 1.22

Three different forms of internal bore seen at an interface between two fluids. (a) strength greater than 4. (b) strength between 2 and 4. (c) strength between 1 and 2. From Simpson (1988).

the downstream flow on the density current will be subcritical and will allow the necessary energy loss to occur by the formation of a stationary wave train behind the current head.

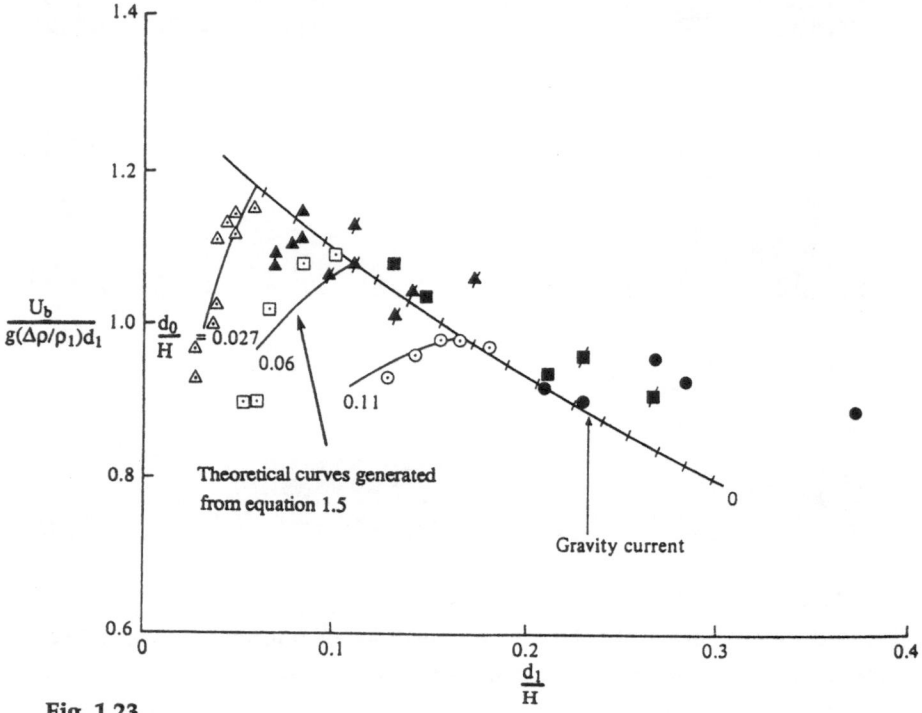

Fig. 1.23

The theoretical curves and experimental points of Wood and Simpson (1984) for a jump advancing into a two layer flow. Modified after Wood and Simpson (*op cit*).

1.8 INTERNAL SOLITARY WAVES: THEORY AND EXPERIMENTAL STUDIES

1.8.1. Introduction

Solitary waves are an important non-linear, permanent waveform that can occur both on the surface of a fluid and within a density stratified medium. The permanence of form results from a constancy of speed and shape that occurs as a result of a balance between non-linear and dispersive effects - a fact that was recognized in the early studies of internal waves (eg. Korteweg and de Vries 1895). Non-linear effects dominate very long waves and result in the propagation speed of the waveform being dependent on the amplitude of the disturbance, which then causes one side of the waveform to steepen during propagation (Kubota *et al* 1978). Dispersive effects, on the other hand, dominate small waves in which the propagation speed is essentially dependent upon the wavelength of the disturbance. This tends to result in spreading out of the waves as they propagate (Kubota *et al* 1978). A solitary wave therefore, represents a balance between non-linear and dispersive effects which enables it to maintain its characteristic permanent form and constant velocity (Kubota *et al* 1978).

Natural internal solitary waves are most commonly observed as occurring along the thermocline in deep lakes, (eg. Thorpe 1971; Farmer 1978) and in the ocean where they can have amplitudes of 100m or more and velocities of a few m s^{-1} (eg. Mysak 1984 and Ostrovsky and Stepanyants 1989). More recently, internal solitary waves have been observed as important features of atmospheric flow (Smith 1988).

1.8.2   Surface solitary waves

The solitary wave was originally described by Russell (1838) as "a large solitary elevation, a rounded, smooth and well-defined heap of water, which continued its course along the channel apparently without change of form or diminution of speed". The discovery and initial work upon this waveform resulted in the the surface solitary wave becoming the subject of both theoretical and experimental study during the mid-nineteenth century (see Miles 1980 for review). An important feature in the subsequent development of internal solitary wave theory was the derivation of the Korteweg and de Vries (KdV) equation for long finite amplitude surface waves, which is commonly expressed thus (Le Blond and Mysak 1978):

$$\eta_t + c_0\eta_x + c_1\eta\eta_x + c_2\eta_{xxx} = 0 \qquad\qquad (1.6)$$

where η is the displacement on the surface of the wave (see Fig. 1.24), $\eta_t = d\eta/dt$, $\eta_x = d\eta/dx$, $\eta_{xxx} = d^3\eta/dx^3$ and,

$$c_0 = (gH)^{1/2} \qquad\qquad (1.7)$$
$$c_1 = 3c_0 / 2H \qquad\qquad \text{and} (1.8)$$
$$c_2 = c_0H^2 / 6 \qquad\qquad (1.9)$$

where g is the acceleration due to gravity, and H is the quiescent water depth.

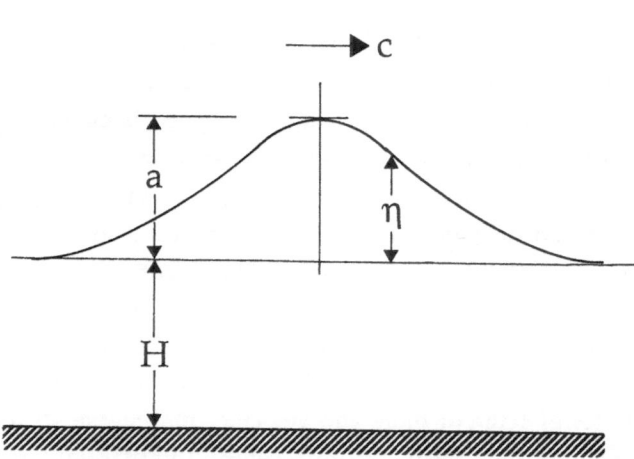

Fig. 1.24

The nomenclature for solitary waves.

The equation that Korteweg and de Vries (1895) originally derived, was more general and included periodic solutions for cnoidal waves. In the limit of long wavelengths, solutions reduced to a solitary wave solution:

$$\eta = a \, \text{sech}^2 \left[\frac{(x - ct)}{L} \right] \tag{1.10}$$

where h is the displacement above the quiescent water depth, a is the amplitude of the wave, x is a horizontal co-ordinate, t represents time and:

$$c = c_0 + \frac{ac_1}{3} = c_0 \left(1 + \frac{a}{2H} \right) \tag{1.11}$$

and, $$L = \left(\frac{12c_2}{ac_1} \right)^{1/2} = \left(\frac{4H^3}{3a} \right)^{1/2} \tag{1.12}$$

From Figure 1.24 it can be seen that L, which is a measure of the horizontal extent of the wave ("wavelength"), corresponds to the width measured from the crest to half the amplitude. Equation (1.12) implies that for any particular depth, the wavelength (L) is proportional to the square root of the amplitude (Mysak 1984).

If, when two such waveforms collide, they emerge with no apparent change of form or velocity and have apparently merely undergone a phase change, the solitary waves are described as solitons (Zabusky and Kruskal 1965). Meiss and Pereira (1978) showed that during collision of solitons the larger wave underwent a positive phase shift whilst the smaller suffered a negative shift.

1.8.3   Internal solitary waves

Detailed study of internal solitary waves did not begin until the mid-twentieth century, when study was based upon the extrapolation of the surface solitary wave theory of Korteweg and de Vries (1895) to explore the possibility of shallow water internal solitary wave propagation (eg. Keulegan 1953; Long 1956; Benjamin 1966). Study of internal waves proliferated with considerations of propagation of internal solitary waves in deep water (eg. Benjamin 1967; Ono 1975), and in a finite depth fluid (eg. Joseph 1977). Experimental work concentrated on comparisons of laboratory scale internal waves with theory (eg. Davis and Acrivos 1967, Koop and Butler 1981, Segur and Hammack 1982) and measurements of solitary waves in the environment (eg. Thorpe 1971,Thorpe et al 1972; Ostrovsky and Stepanyants 1989). Recently, study has expanded to consider the effects due to a variety of factors that are not considered in the basic evolutionary equations. These include propagation of internal solitary waves in a shear flow (Thorpe 1978a, b) and also in a channel of arbitrary cross section (Grimshaw 1978). The causes of dissipation of solitary waves is another area in which work has concentrated (see Grimshaw 1983 for a review of this work).

Koop and Butler (1981), in a detailed comparison of theoretical and experimental results, categorised the existing analytical work as follows:

(i) Shallow water theory:
$$L_{KDV} / H \gg 1, \qquad\qquad h /H = O\ (1) \qquad\qquad (1.13)$$
with the scaling of the wave being such that : $L_{KDV}/H = O\ (a/H)^{-1/2}$ and L_{KDV} = the horizontal extent of the wave in the shallow water internal solitary wave theory.

(ii) Deep water theory:
$$L_{BDO} / H \longrightarrow 0 \qquad\qquad L_{BDO} / h \gg 1 \qquad\qquad (1.14)$$
where the scaling of the wave is $L_{BDO} / h = O\ (a/h)^{-1}$ and L_{BDO} = the horizontal extent of the wave in the deep water internal solitary wave theory.

(iii) Finite depth theory:
$$L_{FD} / h \gg 1 \qquad\qquad h /H \ll 1 \qquad\qquad (1.15)$$

and L_{FD} = the horizontal extent of the wave in the finite depth internal solitary wave theory and where L is a measure of the horizontal extent of the wave, h is an intrinsic length scale that is taken to represent the thickness over which the density change occurs and H is the total fluid depth.

The Boussinesq approximation ignores inertial effects that result from density variations and provides a useful simplification that has been used in the majority of theoretical examinations of internal solitary waves. However, both Long (1965) and Benjamin (1966) point out that it may be an invalid simplification as it could result in the suppression of some solitary wave solutions.

1.8.4 Shallow water theory

The theory of propagation of shallow water internal waves is based upon the assumption that the wavelength of the waves is long when compared to the total depth of fluid. Keulegan (1953) theoretically derived first order equations for the velocity of propagation of an internal solitary wave between a lower dense layer of thickness d_0 and density ρ_1 that was stably overlain by a less dense layer of thickness d_0' and density ρ_2 (Fig. 1.25). It is assumed for the purpose of simplification that $\rho_1 - \rho_2 = \Delta\rho / \rho_1$ and that $\rho_1 - \rho_2$ had a very small value.

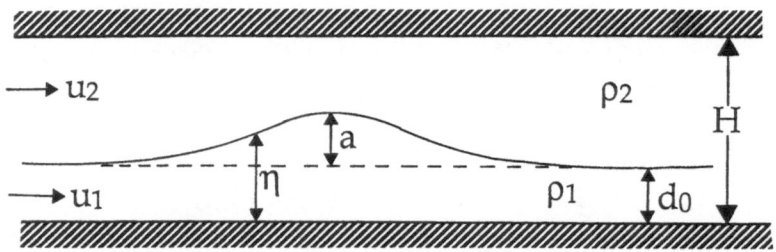

Fig. 1.25
The nomenclature for internal solitary waves. The upper boundary may be fixed or free.

The upper layer was bounded by air at atmospheric pressure and the lower layer by a horizontal bed. Both liquids were initially at rest (ie. $u_1 = u_2 = 0$, where u_1 and u_2 are shear flows in the lower and upper layers respectively) therefore, resulting in irrotational flow. Vorticity was ignored and the velocity of the internal solitary wave was considered to vary in the x-direction only. No reference was made to the fluid depths over which the theory was valid, except to state that for the theory to be applicable, then h << H. The velocity of a solitary wave was then derived and can be represented thus:

$$c = c_0' \left[1 + \left(\frac{d_0' - d_0}{d_0 d_0'} \right) a \right]^{1/2} \tag{1.16}$$

where, $c_0' = \left(\frac{g\Delta\rho}{\rho_2} \cdot \frac{d_0 d_0'}{H} \right)^{1/2}$ (1.17)

and the profile of a solitary wave is:

$$\eta = a \operatorname{sech}^2 \alpha \left(\frac{x - ct}{d_0} \right) \tag{1.18}$$

where c was the velocity of propagation of the waveform, c_0 was the free surface velocity of the waveform, η was the position on the wave profile, x was the horizontal position on the profile, a was the amplitude of the waveform and,

$$\alpha = \left(\frac{a}{2d_0} \cdot \frac{d_0' - d_0}{d_0'} \right)^{1/2} \tag{1.19}$$

Keulegan's considerations indicated that when $d_0' > d_0$ the internal solitary wave is a wave of elevation and when $d_0 > d_0'$ the wave is one of depression. For the situation of $d_0 = d_0'$, the analysis of Keulegan (1953) implied the formation of internal solitary waves to be impossible.

Long (1956) considered the propagation of internal solitary waves on the interface between two incompressible, frictionless fluids of different density that were bounded by rigid boundaries. Long (1956) described a more general case than Keulegan (1953) as the initial velocity of the lower (u_1) and upper (u_2) fluids was not zero and the velocity of the two layers did not necessarily have to be equal. Long (1956) found that a wave of elevation was favoured by: (i) a large density difference, (ii) a high velocity in the upper layer in the direction of wave propagation and (iii) a shallow lower layer. Long (*op cit*) gave the profile of the solitary wave as:

$$\eta = \frac{\beta \operatorname{sech}^2 x}{2\gamma d_0} \tag{1.20}$$

where $\gamma^2 = \dfrac{F_1{}^2 + \sigma F_2{}^2\ ((1-R)/R)}{3\beta(F_1{}^2 - ((\sigma F_2{}^2 R^2) / (1-R)^2)},$ (1.21)

$$F_1{}^2 = u_1{}^2 / g'd_0, \qquad F_2{}^2 = u_2{}^2 / g'd_0 \qquad (1.22a,b)$$
$$\sigma = \rho_2 / \rho_1 \qquad (1.23)$$
$$g' = (\rho_1 - \rho_2)/\rho_2 \qquad (1.24)$$
$$\beta = a / d_0 \qquad (1.25)$$

and, $R = d_0 / H$ (2.26)

Benjamin (1966) considered the propagation of long internal waves, that had developed on a horizontal stream of incompressible fluid, which had lower and upper fixed boundaries. The velocity and density of the fluid were functions of height. The upper surface was fixed. Benjamin (1966) noted that for internal solitary waves the velocity of the wave must always be supercritical. Considering the propagation of a single solitary wave in a specific two layer system that was initially at rest, Benjamin (1966) derived the following expression for the crestal velocity (c):

$$\frac{c^2}{c'^2} = 1 + \frac{((1 - R)^2 - \sigma R^2)}{(1 - R)(1 - R + \sigma R)} \cdot \frac{a}{d_0} \qquad (1.27)$$

where c was the critical wave velocity such that:

$$c'^2 = \frac{(1 - R)(1 - \sigma)}{1 - R + \sigma R} \cdot g d_0 \qquad (1.28)$$

The displacement of the interface was then expressed as:

$$y(\eta_1) - \eta_1 = a\,\mathrm{sech}^2\, \varepsilon x \qquad (1.29)$$

where, $\varepsilon^2 = \dfrac{3aR((1 - R)^2 - \sigma R^2)}{4d_0{}^3 (1 - R)^2(R + \sigma - \sigma R)} \cdot \dfrac{c'^2}{c^2}$ (1.30)

Results (1.27) to (1.30) agree with those of Keulegan (1953) and Long (1956) for the conditions of $F_1 = F_2$ (ie. where the fluids in which the internal solitary waves propagate were initially at rest). The equations reduce to those of a free surface solitary wave if ρ_2 is made vanishingly small (Benjamin 1966). When the expression $(1 - R)^2 - \sigma R^2 > 0$, the wave is one of elevation and when less than zero one of depression. From this, Benjamin (1967) concluded that when the fractional density difference (defined as $(1 - \sigma)$ = $(\rho_1 - \rho_2) / \rho_1$) was very small, the wave was one of elevation when the upper layer was the deeper of the two and of depression when the lower layer was deeper.

Walker (1973) included the effects of surface tension in his analytical study of the shallow water theory. He derived equations that implied no change in the phase velocity when compared with the earlier theories outlined above. Regardless of whether the upper boundary was fixed or free, Walker's (1973) study, to a first order approximation, suggested that the effects of surface tension produced a waveform with a steeper profile (ie. surface tension reduces the effective wavelength of the waveform).

A dispersion relation describes the ease with which waveforms disperse once generated and is an equality between both the angular frequency ω, wave number k (which is expressed in terms of the phase velocity, $c = \omega/k$ as well as wavelength, $\lambda = 2\pi/k$) (Turner 1973). The linear dispersion law for shallow water, internal solitary waves travelling in the positive direction derived by Benjamin (1966) could be written as follows:

$$c(k) = c_0 (1 - \beta'k^2) \qquad (1.31)$$

where $c_0 = (gd_0)^{1/2}$ and $\beta' = 1/6d_0$. The relation is identical to the shallow water surface wave dispersion equation for an infinitesimal amplitude surface wave (Benjamin (1967)). The KdV equation (equation 1.6) can subsequently be derived from equation (1.31) (see Mysak 1984 for derivation).

1.8.5   Deep water theory

The shallow water theory includes the effects of both the lower and upper boundaries in the theoretical models and as a consequence the derived equations are invalid in deep water situations, in which the effects of the upper boundary are not important. This was illustrated by Benjamin (1967), who demonstrated that when the depth of one of the layers was made infinite, the shallow water theory implied that the wavelength of the waveform would increase without bound. Rather than believe that internal solitary waves would not form in fluids of great depth, Benjamin (1967) suggested that this implied that the method of analysis was invalid for the parameters used.

Kubota et al (1978) stated that the wave dispersion relations represented the fundamental difference between the shallow and deep water theories. The dispersion relation for the propagation of long, internal permanent waves in a fluid of infinite total fluid depth, has a linear dependence on the wave number k and can be expressed as :

$$\omega = kc_0 (1 - \psi |k|) \qquad (1.32)$$

where $\psi > 0$ and is equal to $0.5\,(\rho_2/\rho_1)d_0$. The linear dependence of the wave number indicated that the waves in the deep water system were more dispersive than in the shallow water system in which the wave number had a quadratic dependence (see equation (1.31) and also Koop and Butler 1981). The governing evolution equation that describes the propagation of internal waves in fluids of great depth can be derived from the dispersive relation given by equation (1.32) and is the counterpart to the KdV equation in the shallow water situation (Benjamin 1967):

$$\eta_t + c_0\eta_x + c_1\eta\eta_x + c_2 d\frac{d}{dx}\left(\int_{-\infty}^{\infty}\eta(x',t)\,\frac{1}{2\pi}\left(\int_{-\infty}^{\infty}|k|\,e^{ik(x-x')}\,dk\right)dx\right) = 0 \tag{1.33}$$

where $\eta(x',t)$ is the non-dimensional internal wave displacement field. This expression is known as the Benjamin-Davis-Ono (BDO) equation after the main workers on this theory. For the specific case of two stably stratified fluids, bounded by rigid, horizontal plates, in which a uniform layer of depth d_0 and constant density ρ_1 was beneath a uniform layer of infinite depth and density ρ_2 the solitary wave solutions to the BDO equation are (Benjamin 1967):

$$\eta = \frac{aL^2}{x^2 + L^2} \qquad \text{where } x = x - ct \tag{1.34}$$

which applies with,

$$c^2 = c_0^2\left(1 + \frac{3a}{4h}\right), \tag{1.35}$$

and,

$$L = \frac{4\rho_2.d_0}{3\rho_1\,a} \quad *, \qquad c_0^2 = \frac{(\rho_1 - \rho_2)gd_0}{\rho_1} \tag{1.36}$$

Davis and Acrivos (1967) also published work on the theory of the propagation of internal solitary waves in a fluid of infinite depth. The use of an inviscid analysis and the application of the Boussinesq Approximation reduced their results to a relatively simple form which complemented those of Benjamin (1967). In order to check the theoretical results, internal solitary waves were generated by the sudden disturbance of a saline layer that underlay a freshwater layer in a rectangular flume tank. Comparison of the theoretical work with experimental data suggested that the theory provided a good description of observed internal solitary waves.

More recently, Ono (1975) extended the theory of Benjamin (1967) in order to examine the unsteady behaviour of deep water internal solitary waves. He showed that the solutions of Benjamin (1967) were particular ones, but was unable to prove that the

---

*Error in text of Benjamin (1967) corrected

solutions were those of solitons. Meiss and Pereira (1978) confirmed, numerically, the study of Ono (1975).

1.8.6 Finite depth theory

In a theoretical examination of the propagation of solitary waves in a finite depth fluid, in which the length over which the density change occurred (h) was much less than H, Joseph (1977) derived the dispersion relation as:

$$\frac{c(k) - c_0}{c_0} = 1 \quad \frac{kh}{2}\left(\coth (kH) - \frac{1}{kH}\right) \tag{1.37}$$

In the limits of the shallow and deep water theories equation (1.36) reduces to the dispersion relations of the KdV and BDO equations respectively. The finite-depth, internal solitary wave profile can be derived (Joseph 1977):

$$\eta = a\left(\cosh^2 m(x - ct) + \frac{\sinh [m(x - ct)/m]}{n^2}\right)^{-1} \tag{1.38}$$

where, $an = 2c_0 d_0 / C; \quad m\tan(mD) = n^{-1}$ $\qquad$ (1.38a, b)

and, $C_0 = 3c_0/2d_0$ $\qquad$ (1.39)

and as a result, $an = 4d_0^2 / 3$ $\qquad$ (1.40)

Finally, the derivation of the wave speed, c from c_0 is given by:

$$\frac{c-1}{c_0} = \frac{d_0}{2H}\left[1 - 2mH \cot (2mH)\right] = \frac{d_0}{2H}\left[1 + \frac{H (1 - m^2 n^2)}{n}\right] \tag{1.41}$$

Substitution of m and n into the KdV and BDO internal solitary wave profile equations ((1.10) and (1.33)) indicates that m is equal to $1/L_{KdV}$ in the KdV equation and n is equal to L_{BDO} in the BDO equation. The equivalent wavelength scale for the finite depth equation is expressed thus for the finite depth theory:

$$L_{FD} = \frac{1}{m}\cosh^{-1}\left[\frac{1 + 3(mn)^2}{1 + (mn)^2}\right] \tag{1.42}$$

The profile of the waveform varies between two end-member profiles in the shallow and deep water profile respectively. The shallow water internal solitary wave has a steeper profile than the deep water profile and the finite depth profile is intermediate between these two. It should be noted that both the wavenumber (k) and the wavelength scale (L), are not in an explicit form in the finite depth theory, which contrasts with the algebraic relationships of the shallow and deep water theories (see equations 1.31 and 1.32) and Kubota *et al* (1978)).

1.8.7 Experimental observations of internal solitary waves

One of the first published accounts of experimental work upon internal solitary waves was that of Davis and Acrivos (1967) who used experimental results to back up their analytical theory for the propagation of solitary waves in deep water. The study involved propagation of internal solitary waves within the interface separating brine and freshwater in a rectangular flume tank and showed that internal solitary waves could be produced readily from a wide variety of disturbances. Walker (1973) examined the form of the shallow-water, solitary internal wave and investigated the effects of surface tension both analytically and experimentally. The experimentally measured velocities were found to be approximately 8% less than the calculated velocities - a disparity explained by Walker (*op cit*) as resulting from viscous effects within the density interface along which the internal solitary waves propagate.

Maxworthy (1980) examined the two and three dimensional collapse of a denser region of mixed fluid to its equilibrium state within a less dense fluid. In the two dimensional case, the mixed region of fluid was released into a rectangular channel in which salt solution had been allowed to equilibrate beneath fresh water. When released, the mixed fluid moved along the interface and generated internal solitary waves. These waves were found to consist of the mixed region fluid.

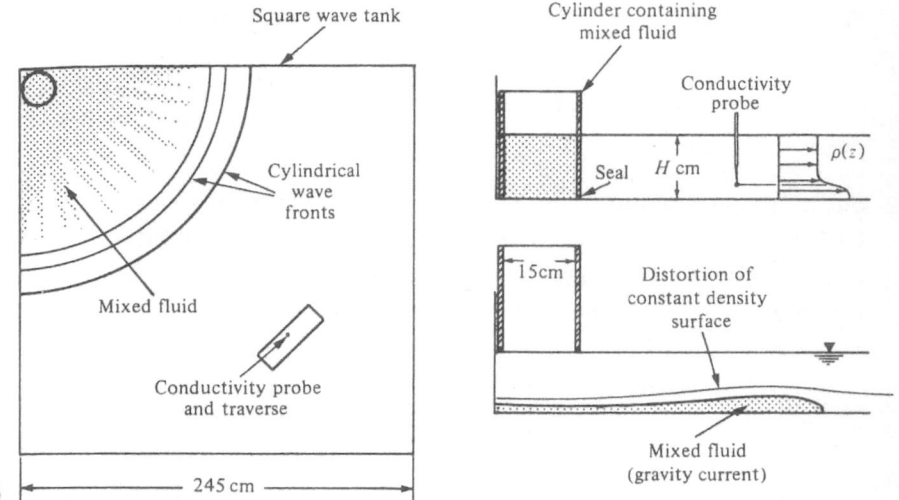

Fig. 1.26

Apparatus to show the collapse of a mixed region in 3-dimensions. From Maxworthy (1980).

Again, with the three dimensional experiments (Fig. 1.26), mixed fluid was released into a tank which had a dense, saline solution basal layer. In this case though, the mixed fluid was more dense than the saline solution and spread as a radial gravity current along the base. The solitary waves broke away from the mixed region front and were again, of the mixed region fluid. Comparison of the generated internal solitary waves with theory determined that they were described adequately by the deep water BDO equation (equation 1.33). The ease with which the experiments produced the solitary waves led Maxworthy (1980) to state that, providing a given physical system was capable of supporting solitary wave motions then, such motions would arise from quite general excitations, ie. that solitary waves were ubiquitous. In the experiments, Maxworthy (1980) observed that the mixed fluid was trapped within the leading solitary wave and ejected rearwards as the wave amplitude decreased due to viscous damping. The generated solitary waves were all ordered by amplitude with the largest amplitude (ie. the fastest) at the front.

An important study was conducted by Koop and Butler (1981) in which a comparison was made between experimentally produced shallow and deep water internal waves and the KdV, BDO and finite depth theories in order to discover the range of conditions over which each theory was valid. They suggested that a fundamental difference between the theories was the amplitude-wavelength scaling. The amplitude-wavelength scaling was also the most sensitive way of establishing the domain of validity for each theory. Their experiments revealed that the wavelength predictions of the two theories differed by more than 100% for a given amplitude and therefore high experimental accuracy was not a pre-requisite. The measured results were found to compare well with the KdV equation for the shallow water theory, although for a/h of less than 0.05 the theory over estimates the velocity by 10 - 15%. This was explained by Koop and Butler (1981) as a result of the attenuation of the amplitude due to viscous effects. The waves generated in order to test the deep water theory did not agree with any of the proposed theories. It appears that the shallow and deep water theories were not valid in the deep water experiments conducted. Furthermore, there was a lack of agreement between the deep water experiments and finite depth theory. In this case, the equation is within its domain of validity and Koop and Butler (1981) could not resolve the problem.

Segur and Hammack (1982) also compared the shallow water and finite depth theories with experimental measurements. They concluded by stating that the KdV equation had a relatively large range of validity and could be used for predictative purposes even when the conditions were satisfied only marginally. However, the finite depth equation had such a small range of validity that it was very difficult to find it, even experimentally.

1.8.8 Other theoretical and experimental considerations

Apart from derivation and examination of equations governing the propagation of internal solitary waves, several workers have considered more general problems relating to the flow of solitary internal waves in the environment. Some of these are outlined below.

(i) Dissipation of internal solitary waves

The most important mechanism of dissipation of internal solitary waves, both in laboratory measurements and within the ocean, is due to viscous boundary layer damping that occurs at the base of the flume tank, on the ocean bed and also within the density change interface. Experimentally, Koop and Butler (1981) found that damping resulted in a 50% attenuation in internal wave amplitude.

(ii) Propagation of internal solitary waves through a stratified shear flow

Various workers have considered the problems associated with the propagation of solitary internal waves through a stratified flow (eg. Benjamin 1966; Benney 1966 and Thorpe 1978a,b). All envisaged a continuously stratified flow except Thorpe (1978a) who considered the effects of shear on the propagation of internal waves along a brine-salt water interface. Thorpe (1978a) demonstrated, both theoretically and experimentally, that increasing the shear flow resulted in the modification of the internal waves. In particular, the effect of shear decreased the wave steepness required in order for breaking to occur thereby, decreasing the range of conditions under which internal solitary waves would form.

(iii) Propagation through a channel of arbitrary cross section

Grimshaw (1978) tackled the problem of changes in solitary internal wave propagation in channels with a variety of cross-sections. The derived equations related the shape of the channel cross-section to the phase speed of the wave. It was found that the equation satisfactorily accounted for velocities of internal solitary waves observed in the shallowing and narrowing Babine Lake, British Columbia (Farmer 1978).

1.9 SUMMARY

(i) Turbidity currents are sediment gravity flows in which the grain suspension is mainly supported by the upward component of fluid turbulence. The currents are an important mechanism in the transport of sediment into the deep ocean.

(ii) Turbidites are the sedimentary deposits of turbidity currents. The sequence of structures is generally believed to represent deposition from a decelerating flow.

(iii) The application of steady state experimental data for sediments transported in clear, fresh water flows to the turbidity currents is being questioned. The effect of sediment fall-out from suspension and the non-steady flow of turbidity currents are considered as additional variables that could explain the deviance of turbidite sequences from what may be expected using present hydrodynamic theories.

(iv) One of the gaps in our understanding of turbidity currents is in the lack of a comprehensive data set for natural turbidity currents. Attempts at modelling real-scale turbidity currents have been unsuccessful. As a result, knowledge regarding the flow of turbidity currents has been derived from experimental modelling and theoretical considerations of both suspension (turbidity) and density currents.

(v) Experimental turbidity current modelling via the finite volume release of a sediment suspension into water has been problematical with only graded beds being generated. However, using sophisticated equipment it has been shown that a Bouma sequence can be generated by the slow deceleration of a sediment suspensions. Nevertheless, the majority of the knowledge of turbidity currents has come from consideration of the flow of density currents. Empirical and theoretical studies have generated relations describing the forward motion of density currents.

(vi) The reflection, or 'ponding', of turbidity currents has been invoked in the past in order to explain certain turbidite sedimentary sequences. However, until recently mechanisms of this ponding had not been considered in detail, it being assumed that a reversed turbidity current would either 'slosh' around in the basin or that another turbidity current would be generated that travelled in the reverse direction. Recently, it was shown that the reversal of a density current from low-angled ramps could generate solitary wave-like forms. This observation opened a previously, largely unexplored avenue of turbidite study.

(vii) Bores are an important natural phenomena that are closely related to density currents with both being important agents for mass transfer. Bores occur as a sudden change in flow depth associated with a hydraulic jump and occur along a density interface. They can be formed by a number of methods including, the intrusion of a density current underneath a layered system.

(viii) Two end-member types of bore can be produced. The first, a strong bore, has many features in common with typical density currents whereas the second, a weak bore, has a smooth upper surface that consists of a train of waves that in time, tend towards solitary waves. Between these two end-members, a continuum of intermediate forms exists with varying amounts of density current and wave characteristics.

(ix) Solitary waves are a non-linear waveform that have a constant speed and permanent shape. The phenomena are important in nature often being associated with bores.

(x) Solitons are a specific type of solitary waves that do not change form or speed when they interact with each other.

(xi) Theories for internal solitary waves have been developed for deep, shallow and finite depth fluids. However, application of these theories to experimental investigations has so far, produced only limited success.

CHAPTER TWO

EXPERIMENTAL STUDY OF THE INCIDENCE OF DENSITY CURRENTS UPON RAMPS AND OBSTACLES.

2.1 INTRODUCTION AND AIMS

This chapter examines, experimentally, the consequences of the incidence of a density current upon various obstructions to its forward flow. In so doing, the contentions of Pantin and Leeder (1987) are considered namely, that the 'reversal' of density currents, and likewise, turbidity currents, results in the generation of a group of internal solitary waves.

The majority of the experiments to be described herein, involved the incidence of restricted 'channelized' density currents upon ramps orientated with strikes perpendicular to the mean flow direction. Consideration was also given to the incidence of: (i) restricted density currents upon ramps, (ii) restricted and unrestricted density currents upon obstacles and (iii) restricted suspension currents upon ramps.

The chapter is structured as follows. After an initial description of the experimental procedure, experimental results are presented which illustrate the effects of the incidence of density currents upon obstructions to the flow path. The experimental results are then compared with theories for the propagation of density currents, bores and internal solitary waves. In conclusion, a model for the flow generated by the reversal, or deflection of density currents from low angled ramps and obstructions is discussed.

2.2 WHY TURBIDITY CURRENTS ARE MODELLED AND THE PROBLEM OF DYNAMIC SIMILARITY

Low density turbidity currents have been observed flowing into lakes and rivers (eg. Gould 1951) but there has been no direct observation of the flow of low to high density turbidity currents into the deep ocean. Information verifying the existence of turbidity currents within the marine environment must therefore, come from the interpretation of data recording instruments and observations of submarine canyons that are scoured by the currents (eg. Inman *et al* 1976; Prior *et al* 1987). However, determination of the physical properties of deep marine turbidity currents is a chancy, expensive and hazardous operation, and a detailed data set has yet to be acquired from the study of an individual natural current.

Alternatively, the physical properties of marine turbidity currents can be determined with the use of a life-sized model. Such an attempt was made by Buffington (1961) who conducted a series of experiments at the head of the La Jolla submarine canyon, Southern California, U.S.A.. Buffington (*op cit*) attempted to initiate high-density turbidity currents by sliding a pre-mixed mass of sediment and sea water onto the seafloor from an elevated, tilted container in water depths of approximately 12 m. Unfortunately, the experiments met with little success, the desired result of generating turbidity currents that entrained sediment and travelled considerable distances not being achieved. The very limited success of real scale experiments has resulted in the major advances in the study of the dynamics of turbidity currents occurring via laboratory and theoretical modelling.

The most straightforward method of modelling density currents is by the sudden release of a volume of denser fluid from a lock, into the remainder of an elongate flume tank, that has been filled with a less dense fluid (eg. Keulegan 1957; Barr 1963; Middleton 1966). When modelling density and turbidity currents in the laboratory, it is important to be certain that the flow patterns observed in the the model are similar to those in the prototype, thereby allowing direct comparisons between the two systems. When this occurs, the model and the prototype are said to be dynamically similar. In a dimensional analysis, Middleton (1966a) stated the following important points concerning the use of models for the steady, uniform flow of both density and turbidity currents:

(i) If the Froude number is similar between the model and the prototype, the two flow types will be similar.

(ii) Attainment of Froude number similarity between the model and the prototype is most easily obtained by having the flow upon a slightly greater incline in the model.

(iii) The velocity scale must be the square root of the length scale. Keulegan (1957) also added that in the case of an unsteady flow, such as a lock-exchange density current, the model timescale must be the square root of the prototype timescale.

(iv) When modelling turbidity currents in which suspended sediment provides the density excess, the ratio of the settling velocity to the average flow velocity must be the same in both the model and the prototype. Thus, the settling velocity will also have to be the square root of the length scale.

2.3   PROCEDURES USED IN THE PRESENT EXPERIMENTS

2.3.1 Apparatus used in the present experiments

The experimental apparatus used by previous workers examining density and turbidity currents, bores and internal solitary waves are given in Tables 2.1 to 2.5.

For the experiments in this study, a simple lock-exchange flume tank was used in order to generate density currents, by the release of a denser fluid into a less dense fluid by the

Author	Tank type	Dimensions l x w x d (m)	Tank orientation: horizontal (h) or inclined (i)	Density current fluid	Ambient fluid
Keuleugan (1957)	Lock-exchange			salt solution	water
Keulegan (1958)	"			"	"
Barr (1963b)	(i) Lock-exchange & re-circulating	5.8 × 0.46 × 0.25	h	more dense water	"
	(ii) Enclosed with central lock-gate	4.9 × 0.1 × 0.1	h	"	"
	(iii) "	0.1 × 0.04 × 0.006	h	"	"
Barr (1963c)	Lock-exchange	87.5 × 0.52 × 0.42	h	sugar solution	"
Middleton (1966 b,c, & 1967)	(i) Re-circulating	5 × 0.15 × 0.5	i	salt solution	"
Simpson (1972)	Lock-exchange	1.5 × 0.3 × 0.3	h	"	"
Kersey and Hsu (1976)	Lock-exchange	5.76 × 0.25 × 0.55	h and i	"	"
Britter and Simpson 1978	Re-circulating	l = 1; w = 12	h	"	"
Simpson and Britter	(i) Re-circulating	l = 1; w = 12	h	"	"
	ii) Modified lock-exchange	1.8 × 0.1 × 0.2			
Britter and Linden 1980	Modified lock-exchange	2.4 × 0.15 × 0.6	i: 0 to 90°	"	"
Huppert and Simpson 1980	(i) Lock-exchange	9.6 × 0.27 × 0.5	h	"	"
	(ii) 12° sector	l = 3.5; d = 0.18	h	"	"
Beghin, Hopfinger & Britter (1981)	Lock-exchange	i) 3 × 0.3 × 0.5	i): 0° to 40°	"	"
		ii) 2 × 0.3 × 0.6	ii) i: 45° to 90°	"	"
Rottman and Simpson 1983	Lock-exchange	3.48 × 0.205 × 0.5	h	"	"
Laval et al (1988)	Lock-exchange	4 × 0.2 × 0.35	h and i	"	"

Table 2.1: Apparatus used by previous workers in order to study the generation of density currents.

Author	Tank type	Dimensions l x w x d (m)	Flume orientation: horizontal (h) or inclined (i)	Denser material	Less dense fluid
Keunen and Migliorini (1950)	(i) Modified lock-exchange	l = 2	h and i	Natural sediment	Water
	(ii) Lock-exchange	31 x 0.6 x 0.6	h and i	Clay	"
Middleton 1967	lock-exchange	5 x 0.15 x 0.5	h	Plastic beads	"
Riddell (1969)	(i) Lock exchange (circular section)	l = 5.72 diameter = 0.14	"	Marble flour	"
	(ii) Lock-exchange	a) 0.508 x 0.1 x 0.1	"	"	"
		b)0.686 x 0.03 x 0.03	"	"	"
Luthi (1980)	Lock-exchange	5.7 x 0.25 x 0.55	h and i	Chalk powder	"
Parker and Fukashima (1987)	Modified lock-exchange	90 x 0.7 x 1.8	i	Crushed silica-Glass	Mississippi river water
Laval et al (1988)	"	4 x 0.2 x 0.35	"	Sand	Water

Table 2.2: Apparatus used by previous workers in order to study the generation of turbidity currents.

Author	Tank type	Dimensions l x w x d	Flume orientation: horizontal (h) or inclined (i)	Denser fluid	Less dense fluid	Ramp (°)
Pantin & Leeder (1987)	Lock-exchange	5 x 0.2 x 0.2	h	Salt solution	Water	7° to 10°
Muck & Underwood 1990	Lock-exchange (?)	1.9 x 0.5 x 0.2	h and i	"	"	28°

Table 2.3: Apparatus used by previous workers in order to study the reversal of density currents

Author	Tank type	Dimensions l x w x d (m)	Tank orientation: horizontal (h) or inclined (i)	Denser fluid	Less dense fluid	Miscible (m) or immiscible (im)
Davis & Acrivos (1967)	Tank	2.5 x 0.1 x 0.4	h	Salt solution	Water	m
Walker (1973)	"	1.8 x 0.54 x 0.12	h	Water	Varsol I	im
Thorpe (1978a)	Rotatable tube	4.85 x 0.1 x 0.16	i	Salt solution	Water	m
Thorpe (1978b)	"	"	h and i	"	"	"
Maxworthy (1980)	(i) Tank (ii) Square tank	8.2 x 0.2 x 0.3 2.45 x 2.45 x 0.1	h "	"	"	"
Koop and Butler (1981)	Tank	6.0 x 0.45 x0.6	"	Freon	"	im
Hammack & Segur 1982	Tank	30.0 x 0.39 x 0.6	"	Salt solution	"	m

Table 2.4: Apparatus used by previous workers in order to examine internal solitary wave formation.

Author	Flume tank	Dimensions l xw x d (m)	Flume orientation: horizontal (h) or inclined (i)	Denser fluid	Less dense fluid	Miscible (m) or immiscible (im)
Wood & Simpson (1984)	(i) Tank (ii) Lock-exchang	3.7 x 0.2 x 0.5 "	h "	Salt solution "	Water "	m "
Baines (1984)	Tank	l = 9.17; w = 0.23	h	Water	Kerosene	im
Rottman & Simpson (1989)	"	3.48 x 0.205 x 0.5	h	Salt solution	Water	m

Table 2.5: Apparatus used by previous workers in order to study the generation of bores.

smooth, rapid removal of a lock gate. The majority of this authors experiments were conducted in a perspex, rectangular lock-exchange flume tank which had a length of 5.0m, a 0.2m square cross section and a barrier 0.325 m from one end (Fig. 2.1). Experiments were also conducted in a perspex tank in which a 1.42 x 0.2 x 0.2m channel, with a barrier 0.25m from the far end, opened out into the centre of one side of a 1m square plan tank which had a depth of 0.2 metres (Fig. 2.2). This tank will be referred to as the "square" tank. Both flume tanks had smooth, horizontal bases.

In order to study the incidence of restricted density currents upon planar inclined surfaces, smooth plastic ramps were inserted into the rectangular flume tank. In the rectangular flume tank, three ramps were used. The first has a length of 1.21m and could be used at ramp angles between 5 and 10° whereas, the other two ramps had lengths of 0.71 and 0.61m and could be used at 15 and 20° respectively. All three ramps had tapered edges, thus reducing the step where the foot of the ramp meets the tank floor to a minimum. When reversals from a 5° ramp were required, it was necessary to tape down the foot of the ramp in order to prevent the flow of minor amounts of the denser fluid under the ramp. All three of the ramps had a close fit with the vertical walls of the flume tank preventing flow of denser fluid between the ramp and the sidewalls of the flume tank.

To study the incidence of an unrestricted density currents upon planar inclined surfaces, ramps of 5°, 10° and 20° were inserted into the square tank. The ramps extended 0.57m along each side and in plan, made an angle of 45° with the axis of symmetry of the spreading, forward flow. As with the rectangular tank, the edges of the ramps were carefully taped preventing fluid flow through gaps between the ramps and the edge of the flume tank. A ramp was also placed with its strike oriented orthogonal to the spreading axis of the density current.

Preliminary experiments were undertaken to consider the incidence of density currents upon obstacles. In the rectangular tank, rectangular obstacles of various heights, that extended the full width of the flume tank were used, whereas in the square tank, a hemispherical obstacle was utilised.

2.3.2 Flow visualisation techniques used

Although a saline solution is clear and colourless, it can be distinguished from freshwater during a density current flow because of the difference in the refractive indices of the two fluids. However, this distinction was not good enough to enable sufficient detail to be gained from video recordings and it was therefore decided to colour the saline solution. Various methods of colouration were tried (eg. milk, food dyes and ink); the most successful method used was the dissolution of a few crystals of potassium permanganate in

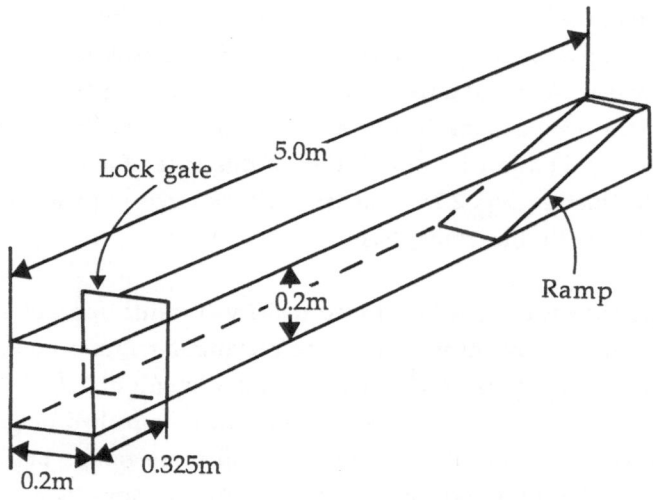

Fig. 2.1
The rectangular tank.

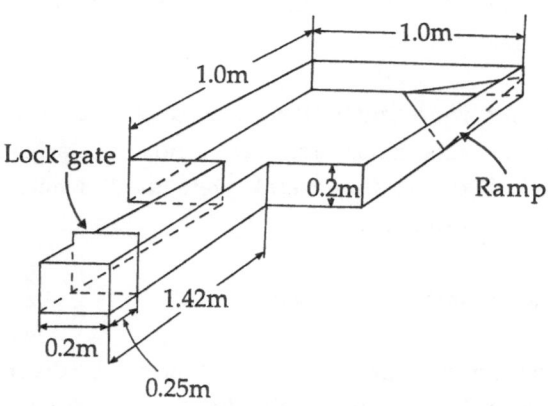

Fig. 2.2
The square tank.

the salt solution prior to its release. Potassium permanganate colours the salt solution a clear, deep purple colour which contrasts well against a white background.

Colouration of the density current enabled accurate measurements to be made regarding two-dimensional flow thicknesses and velocities, but it gave little information about internal fluid motion within the current. After experimentation with a variety of methods, including dye injection and addition of coloured plastic beads to the flow, it was found that utilisation of a low density polymer called Pliolite was most successful (Pliolite is the trade name of polyvinyltoluene butadiene, manufactured by Manchem Ltd, Goodyear Chemicals, U.K.). The Pliolite used was of two densities, 1.050 and 1.546 g cm^{-3}, had an sub-spherical shape with no preferred orientation and a variation in useful grain diameter of approximately 0.5 to 3mm. Pliolite is a highly reflective medium and particle trajectories were most successfully recorded when the flow was lit from above. A little washing-up liquid was added to the saline solution and pliolite mixture to remove any effects resulting from surface tension. It was also found advantageous to soak the Pliolite in water, for at least 24 hours, prior to use in order to remove as much air as possible from the porous polymer.

2.3.3 Flow analysis techniques used

This study has concentrated upon the characteristics of density currents and their reflections as they pass a particular point in the flume tank (ie a Eulerian reference frame). Determination of the flow characteristics from the analysis of high quality, video-recordings, rather than still photography, were preferred since: (i) individual particles could be traced with a high degree of certainty, (ii) velocities could be more accurately determined, as recording times from individual video frames were more precise than those from 35mm stills and, (iii) it was also easier to detect the finer details of motion from video than from still photography.

Individual flows were recorded using a Panasonic F10 camera, U-Matic, high resolution video tape and a portable Sony video-cassette recorder. The video tapes were subsequently analysed using a Sony videocassette recorder (Model VO-5800PS) with a Sony automatic editing control unit (Model RM440) and 18" Sony Trinitron colour monitor. In addition, still shots were taken from the television screen using an Olympus OM-1 SLR camera. These are used for illustrative purposes throughout this chapter.

2.4   SOURCES OF ERROR IN THE EXPERIMENTS

The major sources of error are summarised below along with the action taken to ensure that they were reduced to a minimum. More specific details of the significant errors are given in Appendix 2A.

2.4.1 Errors encountered during experiment set-up

Sources of error in the set-up of the experiments were:

(i) Measurement of the total flow depth

The total flow depth was measured to an accuracy of ±0.5mm (±3% error) and once the saline solution had been mixed the levels either side of the lock-gate were checked again, to make sure that they were still level (to an accuracy of ±0.5mm).

(ii) Variation in temperature between the lock and the main channel.

Errors arising from this were avoided by making the salt solution in the lock, after filling the whole tank to the water depth to be used during the experiment.

(iii) Measurement of the salt content.

Measurement of the salt was accurate to ±10g (±5% error at 200g which decreased to an error of ±0.3% when 3000g of salt were used.

(iv) The initial density of the salt solution.

The initial density of the saline solution was both measured using a density balance (accurate to ±0.0005 cm g^{-3}) and also calculated using density data for the dissolution of NaCl in water at 10, 15 and 20°C (Sohnel and Novotny 1985). Over a 5°C spread in ambient water fluid temperature the calculated densities are accurate to ±0.0005g cm^{-3} at the lowest density used (1.012 g cm^{-3}) which increases to ±0.008g cm^{-3} at the highest density (1.178g cm^{-3}). However, as the temperature variation between experiments was very slight and this was not a problem.

2.4.2 Errors encountered during the experiments

Three potentially significant errors that may result during an individual experiment resulted from:

(i) The non-reproducability of the experiment.

The error resulting from the non-reproducability was estimated by repetition of an experiment and measuring the differences in the flow characteristics (Appendix 2A). This was probably the most significant error encountered during the experiments herein.

(ii) Errors in the recording equipment.

The equipment was maintained in good working order throughout the experiments and gave rise to negligable errors.

(iii) Edge effects.

Boundary layer friction on the sides and base of the flume tank resulted in some local retardation of the flow. The error associated with this could not be quantified. To keep this error to a minimum measurements were made as close as possible to the centre of the flume tank.

2.4.3 Errors encountered during the analysis of the experiments

During analysis of the flow, potential errors arose from:

(i) Spherical aberration of the camera lens and additional distortion by the monitor screen.

This was checked by videoing then subsequently, re-measuring, a centimetre grid on the monitor. There was no measurable distortion, however, as a safeguard, measurements were restricted to the central portion of the cameras field of view.

(ii) Measurement error.

Measurement errors were incurred whilst measuring the property being analysed. As the experiments were enlarged whilst being recorded, measurements could be made with greater accuracy than if from real scale. Details of the significant measurement errors are given in Appendix 2A.

(iii) Analysis of the Pliolite particle paths.

The Pliolite particle paths described a 2-dimensional flow image, although some 3-dimensional movement will have occurred. This will result in some error when velocities are subsequently measured from the particle paths. As the depth of field of the most easily traced Pliolite particles was no more the 3cm, the maximum component of span-wise motion incorporated in the measured particle paths was not large, with the velocities being at most, 5% less than their true values.

2.5   OUTLINE OF EXPERIMENTAL PROGRAMME

2.5.1 Experimental rationale and notation used in the experiments.

There are five controllable variables when considering the incidence of a density current upon a flow obstruction, in a lock-exchange flume tank:

(i) The density difference between the denser fluid and the ambient fluid,

(ii) The ramp angle,

(iii) The position and orientation of the ramp in the flume tank,

(iv) The total flow depth and,

(v) The initial ratio h_0/H, where h_0 is the initial depth of saline solution in the lock and H is the total flow depth.

The major effect of varying the ratio h_0/H will be to change the velocity of advance of the head (U_h) and to provide a range of h_2/H values, where h_2 is the depth of the body (Rottman and Simpson 1983). For purpose of the present study, h_0/H was kept constant at 1, thus giving the maximum value of h_2/H and U_h. Much of the data used in this chapter is derived from the analysis of five series of experiments. In each series a different one of the four remaining variables was kept constant thereby allowing examination of the effects of individual variables.

Experimental conditions are briefly described below and are summarized in Appendix 2B alongside the relevant data derived from the individual experiments. In order to distinguish between individual experiments in the text the following notation is used:

(i) Experiments with constant total flow depth (H).
This includes the S, FE and T series' of experiments. For example, consider experiment FE15/1.053. FE identifies the group of experiments and is arbitrarily chosen; 15 represents the ramp angle (°) and the 1.053 represents the density of the salt solution (g cm^{-3}).

(ii) Experiments with a constant initial density difference ($\Delta\rho$).
This is the notation for the CD series of experiments. For example, consider experiment CD5/7.5. CD is, as above, arbitrarily chosen; 5 is the ramp angle (°) however, 7.5 is the flow depth of the experiment in centimetres.

(iii) Experiments with a constant initial $\Delta\rho$ but with varying observation positions.
This is the notation for the PP series of experiments. For example, consider experiment PP10/102.5. Again, PP is arbitrarily chosen and 10 is the ramp angle. However, the 102.5 refers to the position of the centre of the field of view of the video camera when measured from the end wall of the lock in centimetres (cm).

When, for example, 'the FE10 series of experiments' are referred to, it includes **all** the FE experiments conducted at the 10° ramp angle (see Appendix 2B). Where other experiments are described that are not defined by the notation above, they are clearly annotated.

2.5.2 Summary of experiments conducted

The experiments discussed in this chapter are summarized below. A more detailed description can be found in Appendix 2B.

(i) S5/1.035 to S20/1.178

These experiments examined the flow characteristics of the head of the forward saline density current and the variation in the properties of the reverse flow within the rectangular tank. The foot of the ramp was positioned 270.5cm from the end of the lock and the flow videoed 177.5cm (centre of field of view) from the lock-gate. These experiments utilised the full range of densities (1.012, 1.035, 1.068, 1.100, 1.128, 1.152 and 1.178 g cm^{-3}) and ramp angles (5°, 10°, 15° and 20°). A constant total flow depth of 15.0cm was maintained throughout.

(ii) FE5/1.012 to FE20/1.178

These experiments examined the same properties as experiments S5/1.035 to S20/1.178 but allowed the reverse flow to develop as much as was possible under laboratory conditions. Density currents with initial densities of 1.012, 1.068, 1.128 and 1.178g cm^{-3} were incident upon ramps of 5°, 10°, 15° and 20° in the rectangular flume tank. The foot of each ramp was 370.5cm from the end-wall of the lock and the total fluid depth within the tank was 15.0cm. Two video cameras, centred 50.5cm and 370.5cm from the end wall of the lock were used.

(iii) CD5/5 to CD10/15

In using a constant density (1.068g cm^{-3}), experiments CD5/5 to CD10/15 allowed some consideration of the effects of flow depth on the flow. The flow depth was varied between 5.0 and 15.0cm, in 2.5cm intervals. Two ramp angles (5° and 10°) were used, the positions of both the ramp and the video being as in the S5/1.035 to S20/1.178 series of experiments.

(iv) PP5/102.5 to PP20/209.5

These experiments illustrated the effects of increasing the ramp angle in a flow with both a constant total flow depth (10.0cm) and density (1.053 g cm^{-3}). Ramps of 5°, 10°, 15° and 20° were used, the foot of which were 288.5cm from the end-wall of the lock. The flow was then recorded at 102.5 and 209.5cm from the end-wall of the lock enabling observation of individual waves as they developed with time.

(v) T5/1.065 - T20/1.170

A series of nine experiments using three different densities (1.065, 1.116, 1.170 g cm^{-3}) and three different ramp angles (5°, 10°, 20°) examined the consequences of the incidence of a radially spreading current upon ramps placed in the square tank. The ramp location was kept constant in each experiment, with the foot of the ramp touching the tank sides 57cm from the apex of the ramp, and the dip of the ramp making an angle of 45° with the axis of symmetry of the radially spreading flow. The total flow depth was also kept constant at 10.0cm. An exploratory experiment examined the incidence of an unrestricted density current upon a ramp whose strike was parallel to the sides of the tank opposite to the channel.

(vi) Velocity profile experiments

The experiments aimed to document velocity variations due to turbulence effects at a point as the current passed using Constant Temperature Anemometry (CTA). These experiments had a constant total flow depth of 10.0cm, constant initial density of the saline solution of 1.050g cm^{-3} and a constant ramp angle of 8°; the foot of which was 288.5cm from the end-wall of the lock.

(vii) Head analysis experiment (experiment SC 1)

Using the same conditions as the "Velocity profiles" experiments, this was used for the construction of particle paths for the forward current.

(viii) Sediment suspension experiments

A sediment suspension flow was designed to mimic FE20/1.068, in order to determine whether the effects observed with saline solution density currents could also be produced by turbidity currents.

(ix) Experiments with obstacles

Preliminary experiments examined the consequences of both restricted and unrestricted density currents upon obstacles other than the ramps used above.

2.6   CHARACTERISTICS OF THE FORWARD FLOW

2.6.1 Introduction

Upon the instantaneous release of a denser fluid into a less dense fluid, a surge is produced which rapidly evolves into a current, with a head characterised by lobes, clefts and billows. Behind the head, the body and the tail of the flow develop (Fig. 1.9). The properties of the evolution of an essentially, uni-directional density current have been studied in detail by several workers (chapter 1.6). As a result, features concerning the development of the forward flow which are discussed in this chapter are those that are pertinent to the subsequent consideration of the reverse flow.

The Reynolds numbers of the forward flows varied from 450 at the lowest density used (1.012g cm^{-3}) to a maximum of 1200 at the highest densities (1.178g cm^{-3}). The lowest Reynolds numbers occurred in experiments CD5/5 and 10/5 where they were approximately 150. Thus, in some of the experiments the viscous forces were probably not negligable and will have affected the velocities of the flows. However, in all the experiments conducted (except CD5.5 and CD10/5), the forward flows had a head with well developed billows, lobes and clefts indicating that a fully turbulent flow was generated.

2.6.2 Velocity profiles of the forward flow

Velocity profiles as a density current passed a point in the rectangular flume tank were procured for the experimental conditions described in section 2.5.2. Unfortunately, Constant Temperature Anemometry (CTA) did not prove to be an accurate method of quantifying the flows because each measured value had a non-constant, unpredictable error associated with it (Appendix 2C). However, bearing this in mind, the results of the CTA still provide a useful, qualitative insight into velocity distributions within the flows.

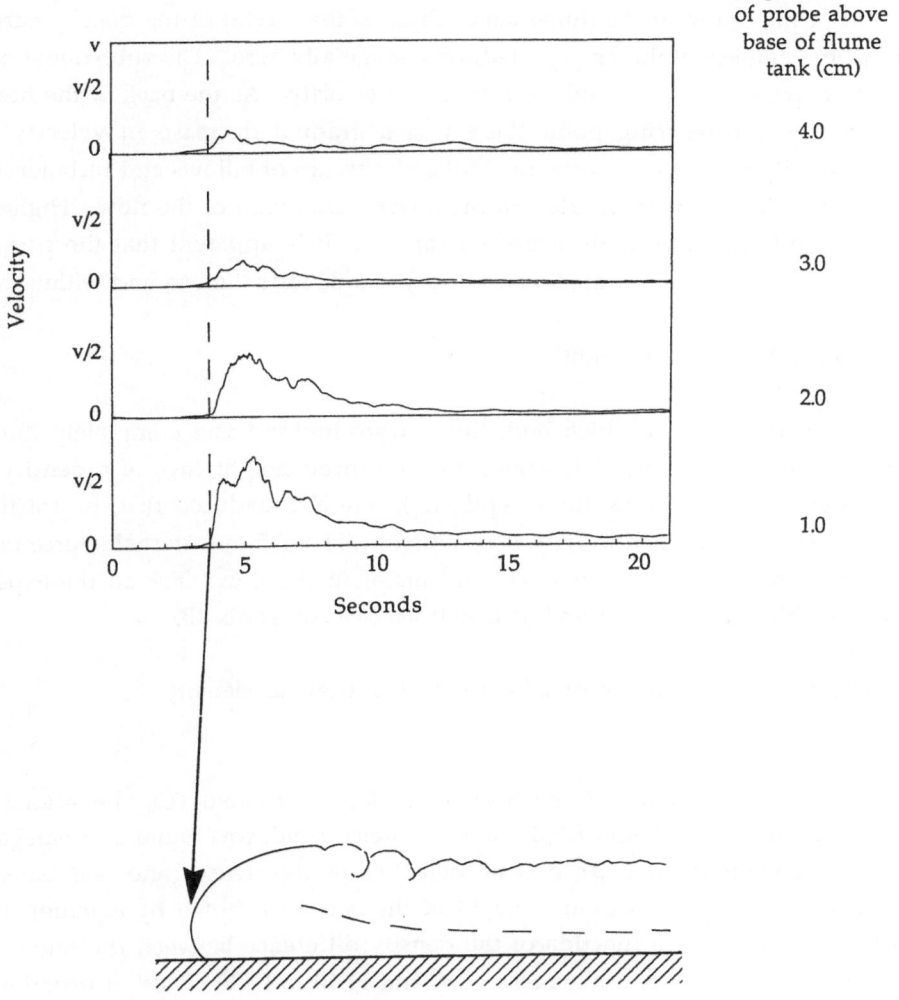

Fig. 2.3

Relative velocity profiles as a density current passes a fixed point in the centre of the rectangular flume tank. The velocity profiles are aligned at the point at which the nose either touches, or is directly below the probe (dashed line). The maximum measured velocity of advance of the head was 10cm s-1. This will be similar to the maximum velocity in the profile measured at 1.0cm above the base of the flume tank.

Figure 2.3 illustrates the variation in 'relative' velocity as the head of the current passes a fixed point in the centre of the flume tank. Prior to the arrival of the density current, the velocity in the ambient fluid, as expected, was essentially zero. The subsequent arrival of the density current incurred a sudden increase in velocity. As the back of the head, body and tail pass the measuring point there was a gradual decrease in velocity. Minor measured velocity fluctuations were due to the occurrence of billows and turbulence. Local instantaneous velocities were greatest in the lowest centimetre of the flow. Higher within the flow, the maximum velocity decreased rapidly. It is apparent that the passage of a density current induces an asymmetrical velocity profile near the bed and within the flow.

2.6.3 Variation in the head height

In a two fluid medium, in which both fluids were inviscid and completely immiscible, Benjamin (1968) theoretically determined that the head height (h_1), of a density current would be exactly half the total fluid depth, (H). He also deduced that, in practice, flow would only be feasible when $h_1/H < 0.5$. When $h_1/H > 0.5$ an external source of energy would be required in order to generate and maintain the flow. For all the experiments conducted by this author, h_1/H was less than 0.500 (see Appendix 2B).

2.6.4 Controls on the velocity of advance of the head of density currents

The velocity of the advance of the head of a density current (U_h) generated by the instantaneous release of a denser fluid into a less dense fluid, was found by Keulegan (1957, 1958) to be dependent upon ρ_1, ρ_2 (the densities of the denser and less dense fluids respectively) and h_1 (the maximum height of the head) as shown by equation 1.4. The measured data for U_h as a function of the density difference between the saline solution and the ambient fresh water for the S5 to S20 and FE10 to FE20 series' of experiments are presented in Figure 2.4.

Initial conditions for all the experimental results presented were identical and therefore, the measured velocities for each particular density should have been the same, as the density was the only variable. It is apparent however, that the velocities of the S5 to S20 series of experiments occurred within a higher range to those of FE5 to FE20 series of experiments. Measurements of U_h for the FE experiments were taken 370.5cm from end-wall of the lock. This distance is equivalent to 11.4 lock lengths and according to Rottman and Simpson (1983), the flow of the density current would no longer be steady and would have just begun to decelerate Conversley, measurement of U_h was made only 5.5 lock lengths (177.5cm) from the end-wall of the lock in the S5 to S20 series of experiments. In this instance, the density currents were still well within their steady phase. Thus the lower range of velocities measured for the FE5 to FE20 series of experiments relative to the S5 to S20 series of experiments is a real effect. In addition, some scatter of velocities occurred within each range. As measurement errors of U_h were very small, the observed variation in the velocities within each range were probably the result of real errors, such as the non-reproducibility of the experiments.

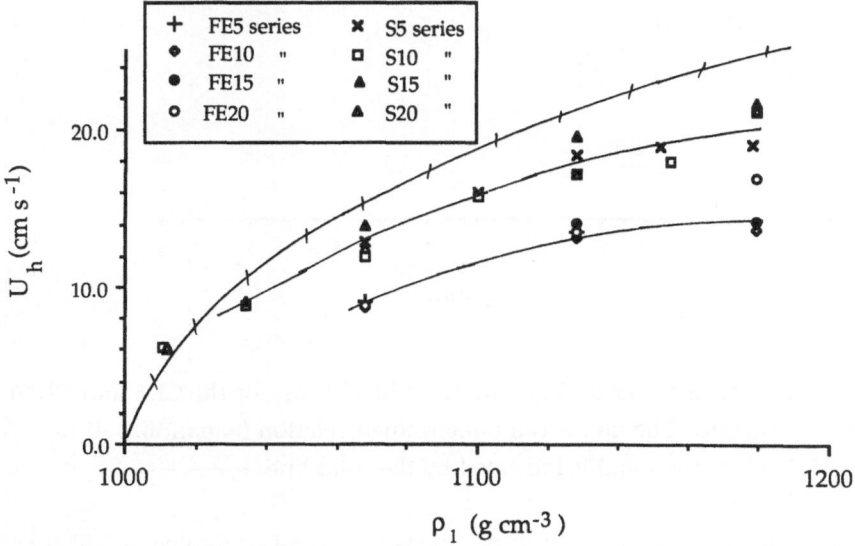

Fig. 2.4
Variation in the velocity of advance of the head (U_h), with the initial density of the saline solution (ρ_1). —⊢— represents the empirically determined relation of Keulegan (1957, 1958, see chapter 1.6, equation 1.4).

The gradient of this trend was less than the equivalent one of Keulegan (1958) - particularly at the highest densities used (1.128 to 1.178 g cm^{-3}). This may reflect the less accurate nature of Keulegan's (*op cit*) velocity measurement techniques, or variation in edge effects between his experiments and those of this author, due to different sized flume tanks.

When a constant density difference was maintained and h_1 varied, U_h was clearly dependent upon h_1 (Fig. 2.5). As in Figure 2.4 ,the relation of Keulegan (1957) predicts higher values of U_h. Some scatter of the values occurs, in this instance, in addition to the error resulting from non-reproducability there is a significant measurement error associated with the measurement of h_1 (see Appendix 2A).

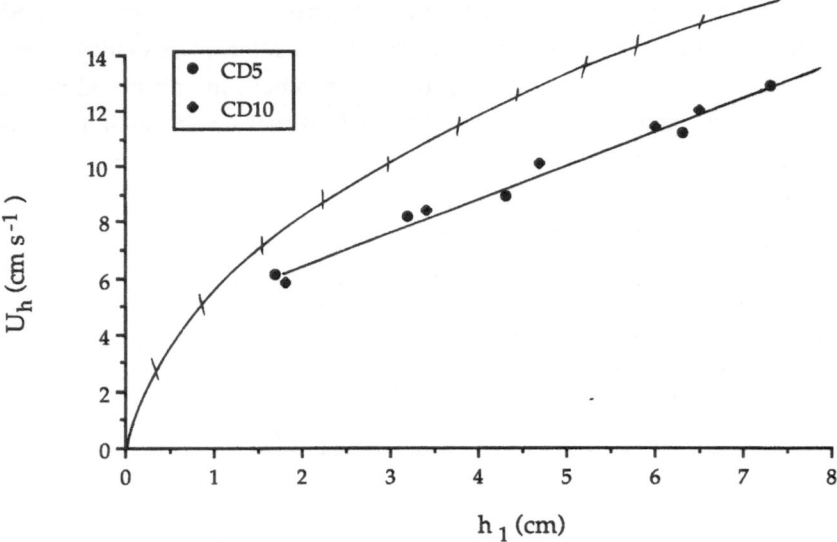

Fig. 2.5
Variation in velocity of advance (U_h) with head height (h_1) for the CD5 and CD10 series of experiments. The empirically determined relation (equation. 1.4) of Keulegan (1957) is plotted and is indicated by the solid line. ⸻⟋⸻⟋⸻

The relationships for the motion of the head of a density current of Keulegan (1957) and of Benjamin (1968) are compared with the measured velocities from this study in Figures 2.6a, b and c.

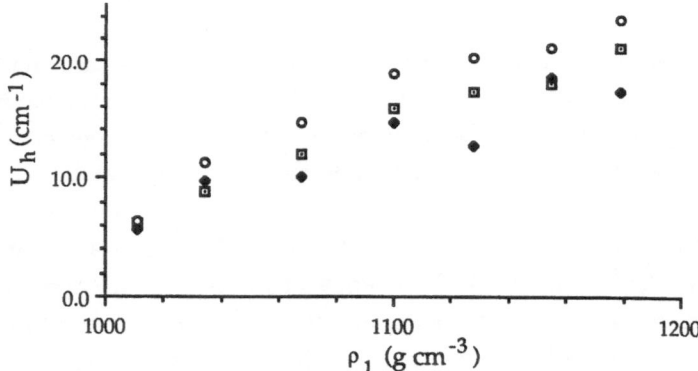

Fig. 2.6a
S10 series of experiments.

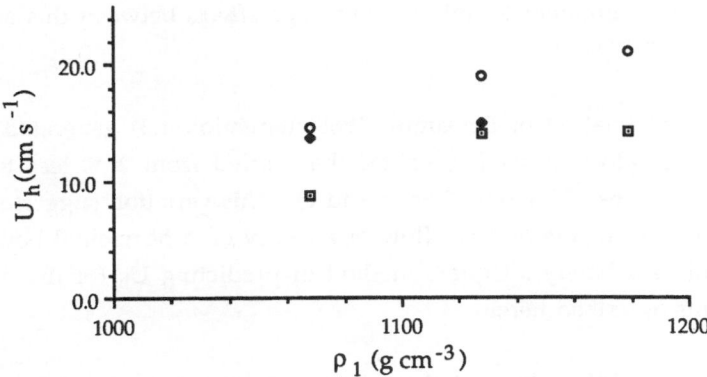

Fig. 2.6b
FE15 series of experiments.

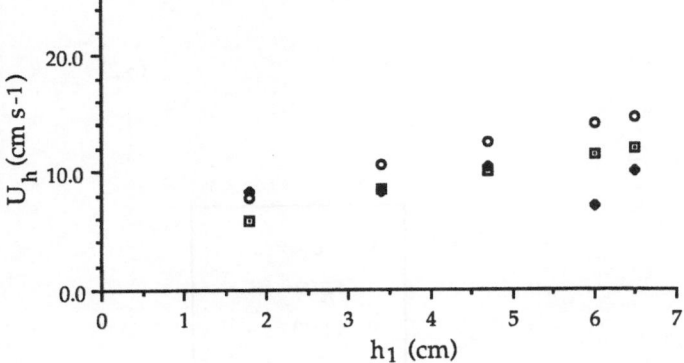

Fig. 2.6c.
CD10 series of experiments.

Fig. 2.6a, b and c.
Comparison of the measured head velocity ⊡, with that predicted by the theories of Keulegan (1957), ○ and Benjamin (1968), ◆.

The empirically determined relation of Keulegan (1957, 1958) generated velocities that were 5 to 56% higher than those that were measured directly from the present experiments. For the data presented in Figure 2.7a, b and c, the greatest discrepancy occurred in the FE15 series of experiments. As discussed previously, this discrepancy was probably largely attributable to the density currents having just entered a stage of deceleration. Also, as mentioned above, the higher values within the S10 series and CD10 series of experiments may reflect both different measurement techniques and edge effects between this authors work and that of Keulegan (1957).

The theoretically determined relation of Benjamin (1968) (equation 1.3) generated some scatter, with the calculated velocities having values that varied from 27% less to 93% greater than the measured values of U_h (Fig. 2.6a, b and c). This variation suggested that the theory, modified from that for the inviscid flow of a cavity on a horizontal boundary above a more dense fluid, is a barely adequate method of predicting U_h for the density currents in the experiments described herein.

An alternative method of examining the variation of the velocity of the head is to non-dimensionalize it as a Froude number and then to compare it with h_2/H, the fractional depth of the body. Two Froude numbers were used in this study (modified after Britter and Simpson 1978):

$$Fr_{h_2} = U_h/(g'h_2)^{1/2} \tag{3.1}$$

$$Fr_H = U_h/(g'H)^{1/2} \tag{3.2}$$

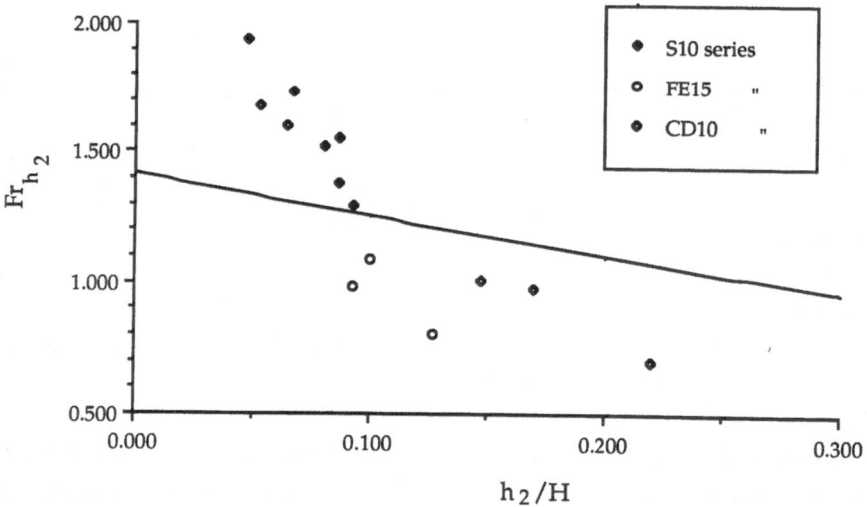

Fig. 2.7 a

Variation of Fr_{h_2} with fractional depth where ——— represents Froude numbers calculated for an inviscid density current flow, with no mixing (Benjamin 1968).

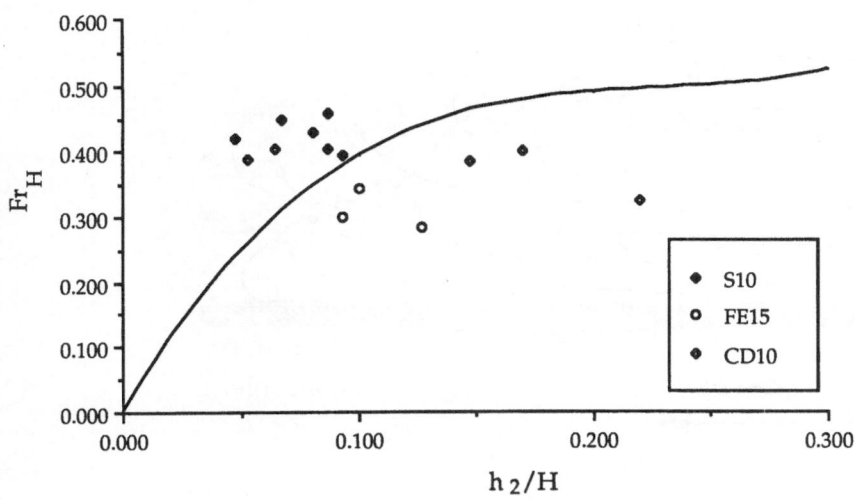

Fig. 2.7 b

Variation of Fr_H with fractional depth where ———— represents the Froude numbers for the inviscid flow of a density current, with no mixing (Benjamin 1968).

in which $g' = g((\rho_1 - \rho_2) / \rho_1)$. Froude numbers calculated from these experiments were then compared with inviscid, non-mixing Froude numbers determined by Benjamin (1968) (Figs. 2.7 a and b)

The related experiments of Britter and Simpson (1978), in which friction on the bottom and sides of the flume tank was zero, showed that mixing at the head of a density current increased the Froude number above those predicted by the theory of Benjamin (1968). For the experiments conducted in this study, this conclusion held at lower values of h_2/H, although the Froude numbers were smaller than those in Britter and Simpson (*op cit*), for equivalent values of h_2/H. However, at higher values of h_2/H, the predicted Froude numbers fell below those suggested by Benjamin's (*op cit*) theory. So, although mixing served to increase Froude numbers relative to inviscid theory, it is apparent that friction on the bottom of the current decreased them, as friction decreased the value of U_h, thereby generating the variation observed in Figures 2.7a and b.

Motion within the central part of the head was largely horizontal (Fig 2.8). At the front of the head, the flow was pushed back and up from the nose and was subsequently caught up in the billows that occurred on the back of the head. Such features have previously been described by Middleton (1966b). Velocities of individual Pliolite particles within the head

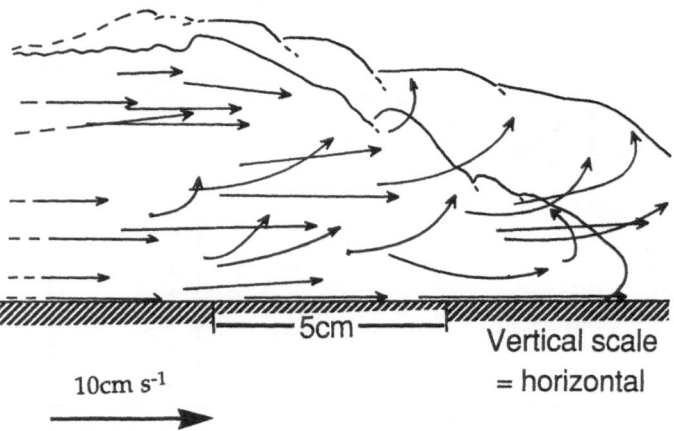

Fig. 2.8
Internal motion within the head of a density current determined from the movement of Pliolite particles (experiment SC1).

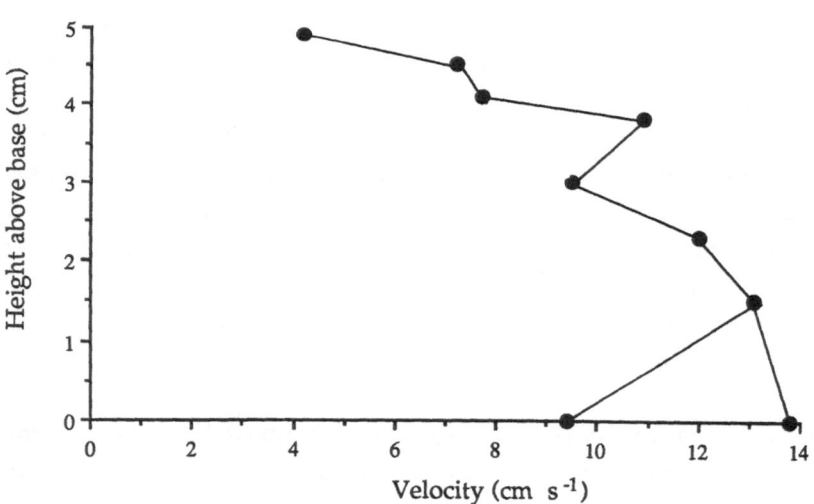

Fig. 2.9
Vertical velocity profile through the central part of the head. The velocities are measured from Pliolite particle paths, with the assumption of 2-dimensional flow.

of Figure 2.9 are presented as a profile in Figure 2.9. The velocities within the lower part of the head are greater than U_h (10.0cm s^{-1}), a fact that had been previously noted by earlier workers (eg. Middleton 1966b; Simpson and Britter 1979). In this instance, the maximum velocity within the head was 38% greater than U_h. This value was 100% greater than that determined by Simpson and Britter (*op cit*), the discrepancy possibly arising from their use of an arrested saline density current for their studies.

2.6.5 Variation of the body height behind the head

The body height is related to the total volume flux of the current which in turn, is a function of the volume of the current initially in the lock. The experiments described herein had a constant total volume flux, as the ratio h_0/H remained constant at 1. This resulted in only a small range of non-dimensionalized tail thicknesses such that, h_2/H varied from 0.047 in S10/1.128 to 0.170 in CD10/10, the variation being a consequence of the non-reproducability of the experiments.

2.6.6 Flow within the tail of the forward flow

The flow velocity rapidly decreased behind the head as the current dissipated energy by frictional damping and mixing. Motion within the tail, as depicted by the movement of Pliolite particles, is illustrated in Figure 2.10. The dominant particle movement direction was parallel to the bed and velocities within the tail varied from, 1.7cm s^{-1} near the base to 7.8cm s^{-1} within the mixing zone (Fig 2.11). The velocities near the base were approximately 12% of the particle velocities near the base of the head (Fig 2.8) whereas, within the mixing zone, particle velocities were 60 to 82% of those within the central part of the head.

2.7 NATURE OF THE CURRENT GENERATED BY THE INCIDENCE OF A RESTRICTED DENSITY CURRENT UPON A RAMP

2.7.1 Introduction

Until the exploratory experiments of Pantin and Leeder (1987), the reversal of a density (or turbidity) current from a ramp was believed to generate either: (i) a turbidity current moving in the reverse direction (eg. Allen 1985) or, (ii) a "sloshing" of sediment suspension in the basin, perhaps with the additional association of internal seiches (eg. Van Andel and Komar 1969). Reflection of a density current from a vertical wall was known to generate internal solitary wave-like forms (eg. Simpson 1988). However, it was not until the work of Pantin and Leeder (*op cit*) that the possibility for the generation of

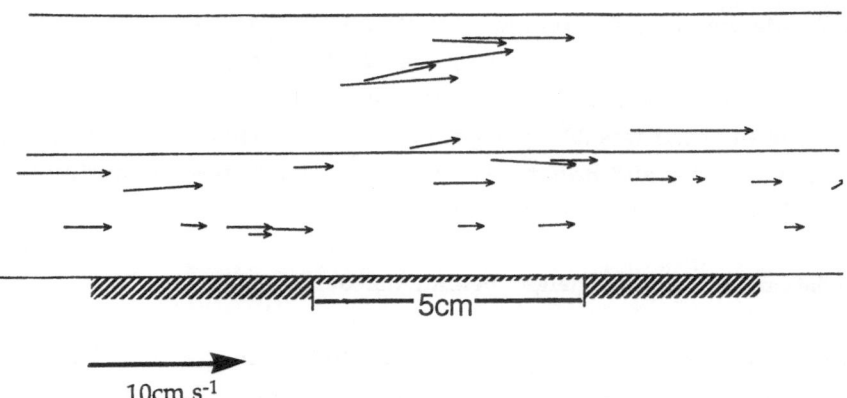

Fig. 2.10
Internal motion within the tail of the density current of experiment SC1 determined from the movement of Pliolte particles.

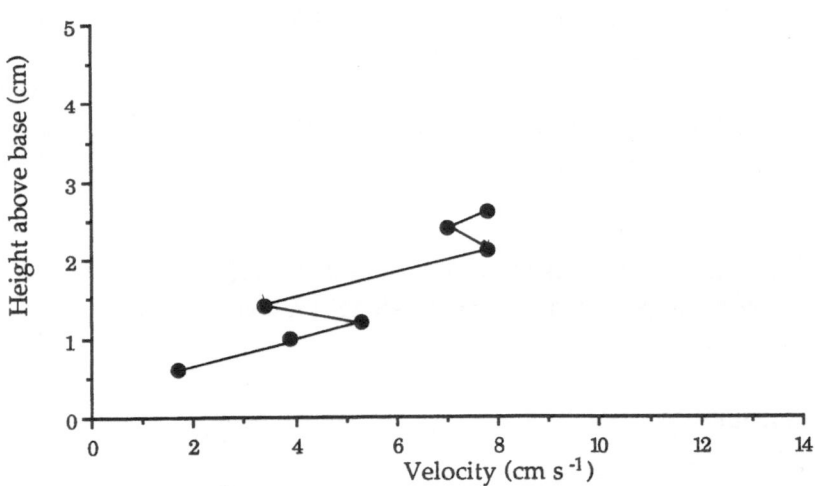

Fig. 2.11
Vertical velocity profile through the tail of the flow. Velocities are measured from Pliolite particle paths, with the assumption of 2-dimensional flow.

internal solitary waves as the direct consequence of the reversal of density (and therefore turbidity) currents was suggested.

The following sections discuss the results of experiments conducted by this author, that examined the consequences of the incidence of both restricted density and turbidity currents upon planar ramps. Initially, individual experiments S15/1.068 and FE15/1.068 will be described, as they illustrate the range of features to be discussed in subsequent sections.

As outlined in section 2.5.2, experiments FE15/1.068 and S15/1.068 involved the reversal of density currents with an initial density of 1.068 g cm^{-3} from a 15° ramp. The differences between the experiments were:

(i) that the foot of the ramp in FE15/1.068 was 370.5cm. from the end-wall of the lock compared with 270.5cm. in S15/1.068 and,

(ii) that the video camera was centred 50.5cm from the end-wall of the lock in FE15/1.068 whereas it was 177.5cm. in S15/1.068.

As discussed in section 2.6.4, these differences in experimental conditions affected the resultant forward flow of the density currents in the following manner. Firstly, the velocity at the foot of the ramp was less in FE15/1.068 (8.8cm s^{-1}) than S15/1.068 (approximately 13.9cm s^{-1}) because in FE15/1.068 frictional damping had been operative for longer and the current had begun to decelerate. Secondly, flow thicknesses within the respective density current heads were different, a consequence of the decrease in velocity. For example, h_1 = 7.1cm in S15/1.068 and 6.5cm in FE15/1.068; h_2/H = 0.060 in S15/1.068 and 0.127 in FE15/1.068. The value of h_2/H in FE15/1.068 was 212% greater than in S15/1.068 possibly because h_2 in FE15/1.068 was made when the head had just begun to travel up the ramp which may have had some influence on the thickness of h_2.

Despite these differences, the two flows generally represent, different developmental stages of the same density current. Such a simplification is adequate for the predominantly, qualitative description of the flow that follows.

2.7.2 Incidence of the density currents in experiments FE15/1.068 and S15/1.068 on ramps of 15°

As described in the preceding sections, the release of a finite volume of saline solution (ρ_1 = 1.068 g cm^{-3}) into fresh water (ρ_1 = 0.999g cm^{-3}) generated a surge which rapidly developed into a density current with a head, body and tail (Plate 2.1). When the density current was 177.5cm from the end-wall of the flume tank, the head of the density current of experiment S15/1.068 had a U_h of 13.9cm s^{-1}, h_1 of 7.1cm, h_2 of 0.9cm and h_3 behind the head of 3.8cm.

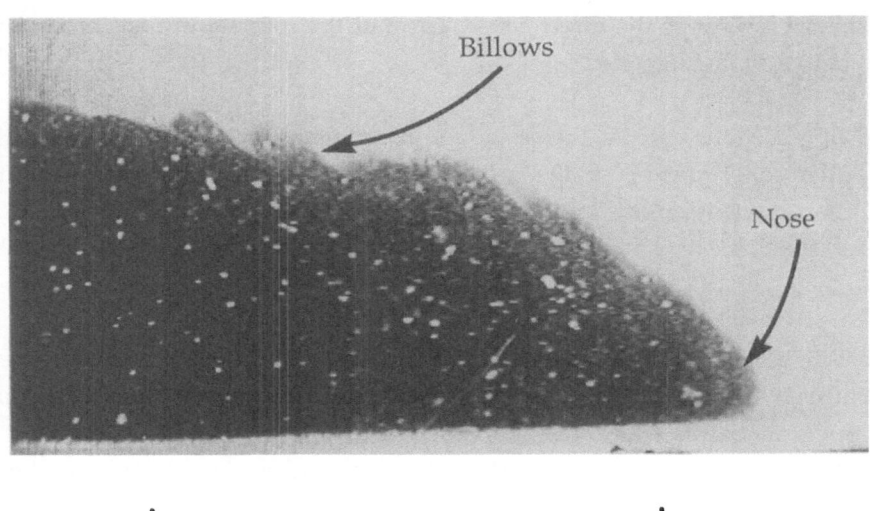

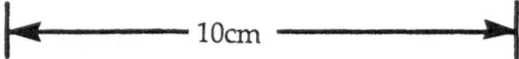

Plate 2.1

Vertical view of the head of a forward-moving, surge style density current.

At this stage Fr_{h_2} was 1.8. As it propagated further, U_h began to decline and thicknesses within the current changed. When the current was 370.5cm from the end wall of the flume in experiment FE15/1.068, U_h was 8.8cm.s^{-1} (37% less than 177.5cm from the end-wall of the lock), h_1 was 6.5cm, h_2 was 1.9cm and h_3 was 4.6cm.

As the head of the density current began to move up the ramp in experiment FE15/1.068 it decelerated continuously before momentarily, coming to a standstill. At this point, the nose of the current was approximately 10cm above the base of the flume tank. The bulk of the body of the flow did not travel as far up the ramp, and instead, accumulated momentarily, near the base of the ramp forming a 'bulge' of fluid. This elevated mass of fluid then began to propagate, through the body and mixing zone of the residual, forward-moving tail of the initial flow, back towards the lock.

By 93cm from the foot of the ramp in experiment S15/1.068, the elevated mass of fluid had degenerated into a two 'hump' phenomenon. The first 'hump' had all the characteristic features of a density current, ie. a strong asymmetry of form with billows, lobes and clefts (Plate 2.2a). The velocity of advance of the front of this reverse flow (U_b) was 10.8cm s^{-1}. Behind the billows on the back of the first 'hump' there was a diffuse region of mixed fluid. This was followed by a poorly defined 'waveform' in which no billows were observed - although pervasive mixing of the saline solution with the fresh water may have masked their existence. The body thickness of the reverse flow behind these elevated masses of fluid was thicker than prior to the arrival of the reverse flow.

Before the arrival of the reverse flow, Pliolite particles within the 0.8cm thick tail were moving parallel to the bed in a forward direction, at velocities that varied between 1.3 and 4.9cm s^{-1}. Within the mixing zone the velocity was also approximately 4.9cm s^{-1} (Fig 2.12a). As the reverse flow propagated through the tail of the forward flow, Pliolite particles near the base of the flow experienced a small push, in the reverse direction.

The local velocity within the first elevated fluid mass was bed-parallel with a maximum of 12.7cm s^{-1} near the base and 11.3cm s^{-1} within the central portion of the 'hump' of fluid (Fig. 2.12 b). The billows generated more complicated flow paths and smaller horizontal displacements than within the central part of the wave. Within the mixing zone, above the 'hump' of denser fluid, the flow still moved in the forward direction although, it was

Fig. 2.12 a to f
Variation in the internal fluid motion before (a and d), during (b and e) and after (c and f) the arrival of the reverse flow front in experiments S15/1.068 and FE15/1.068. The arrows indicate Pliolite particle paths within the flow. The solid line represents the salt solution - fresh water interface at t = 0 seconds and the dashed line the interface at t = 0.32 seconds.

76

Fig. 2.12

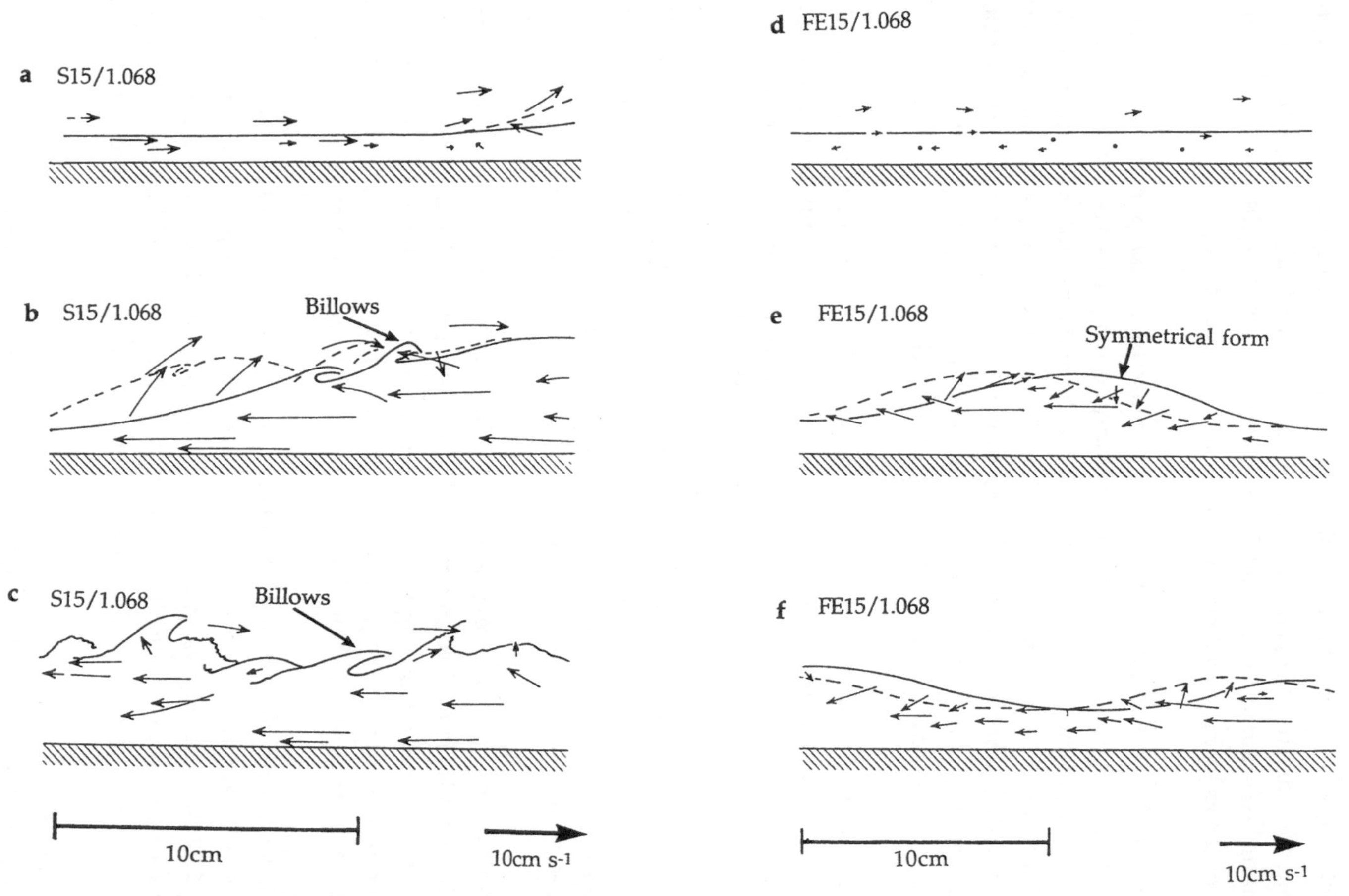

a S15/1.068

b S15/1.068

Billows

c S15/1.068

Billows

10cm

10cm s⁻¹

d FE15/1.068

e FE15/1.068

Symmetrical form

f FE15/1.068

10cm

10cm s⁻¹

Plate 2.2
(overleaf)

Vertical views of the three types of internal bore that are generated by the incidence of a density current upon low angled ramps. The arrows indicate the direction of movement of the bore.

(a) Type A bore.

Characteristic asymmetrical form and billows that are usually associated with the head of a forward-moving, surge style density current.

(b) Type B bore.

Flow separation occurs downstream of the crest of the wave. Flow properties are intermediate between those of a type A and a type B bore.

(c) Type C bore.

The upper surface consists of a group of smooth, approximately symmetrical wave forms.

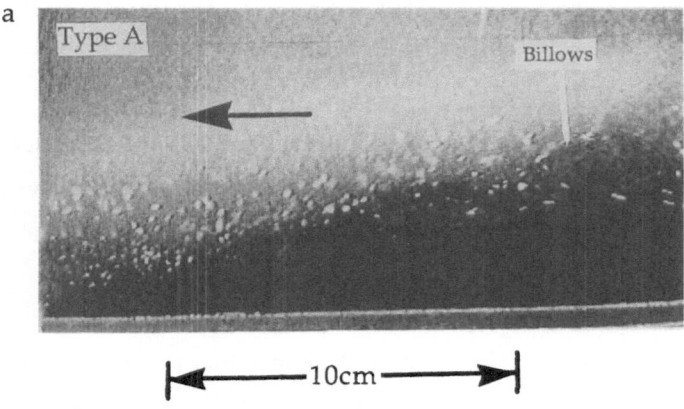

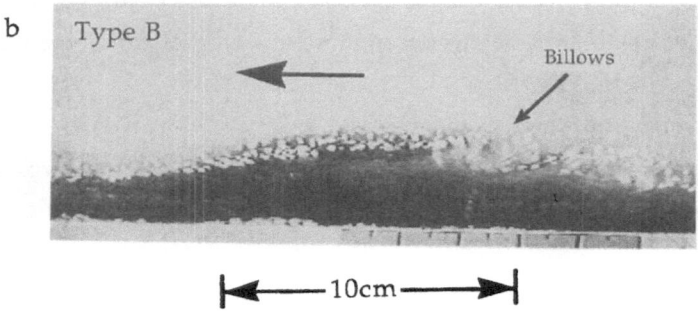

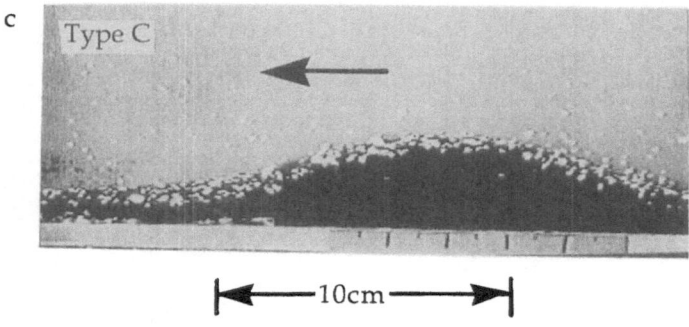

Plate 2.2

deflected by the undercutting reverse flow. The bed-parallel component of the velocities within the mixing region, were comparable to the velocities within in the mixing region prior to the arrival of the reverse flow. In addition, as each 'hump' of fluid cut under the mixing zone, there was a component of motion vertical to the direction of reverse flow motion.

Between the first 'hump' and the following diffuse, elevated mass of fluid, the motion was predominantly horizontal, although in the upper centimetre where the flow was most diffuse and being actively mixed, a vertical component of movement was also important (Fig. 2.12c). Near the base of the flow behind the first 'hump' of denser fluid, the maximum velocity was 9.8cm s^{-1} whilst in the upper part of the flow, the maximum velocity was 1.4cm s^{-1} (Fig. 2.12b).

The reverse flow, 320cm from the foot of the ramp, consisted of a group of four smooth, symmetrical and well-defined waveforms (Plate 2.2c) that were followed by several, poorly defined waveforms. The crestal velocities of the first three were, 10.7, 10.2 and 10.2cm s^{-1}, which were 16 to 22% greater than the velocity of the head of the current as it hit the ramp.

Prior to the arrival of the main current the instantaneous velocities within the tail varied from zero to 1.1cm s^{-1} (Fig. 2.12 d). The direction of movement was bed-parallel and predominantly in the forward direction. However, within the lowest centimetre, there was a slow drift of Pliolite particles in the reverse direction. Above the tail, the velocity was bed-parallel, in the forward direction, with a maximum of 2.0cm s^{-1}.

As the first waveform impinged upon the tail, particles at the front of the wave were pushed along by it. Conversely, particles at the crest of the waveform moved back over the wave, ie. in the forward direction, and those at the back of the wave were pulled along in the reverse direction (Fig. 2.12e). Maximum sub-horizontal displacement, in the reverse direction, occurred within the central part of the wave, where the movement was parallel to the bed, achieving maximum velocities of 10.2cm s^{-1}. Between the first and second waves, motion within the reverse flow remained bed parallel attaining a maximum velocity of 4.6cm s^{-1} (Fig 2.12f).

To summarize, the flow generated by the incidence of a restricted density current upon a 15° ramp consisted of two forms:

(i) a flow which apparently had the characteristics of density currents, ie. an asymmetry of form, billows, lobes and clefts.

(ii) a flow whose upper surface had the profile consisting of a group of waves.

2.7.3 Flow motion at the ramp incurred by the incidence of a density current

As outlined in the previous section, whilst the head of the forward current moved up the ramp, the body of the current accumulated around the foot of the ramp, forming a visible 'bulge'. As soon as the bulge was visible, it began to propagate back towards the lock, as the reverse flow (Plate 2.3).

Considering this sequence in more detail, as the head decelerated whilst moving up the ramp, it thinned rapidly (Plate 2.3 and Figs. 2.13a to i). Associated with this was a thinning of the body of the flow that was also on the ramp. Flow of the current away from the ramp, towards the lock, was clearly discernable in the lower part of the body by 3sec* (Fig. 2.13d). At this point, the bulge was visible and stationary, being supplied with saline fluid from both tail of the forward flow as it approached the ramp and by the downslope flow of fluid from the gravitational collapse of the head. Without the presence of Pliolite tracers, initiation of the reverse flow would not have been apparent until after 5sec., when at which point the 'bulge' began to propagate towards the lock. The bulge moved through all but the uppermost layer of the tail of the the residual forward flow, although as shown by Figures 2.13f to 2.13h, the forward motion in the body had almost been completely eradicated. The nature of the front of the reverse flow is indicated clearly in Figure 2.13g in which the first 'hump' of, for instance, S15/1.068, is present. As the returning flow undercut the residual tail, it entrained the forward moving particles into the reverse motion.

Movement within the mixing region was also affected as the current travelled up the ramp. Initially flow within the billows was suppressed (Fig. 2.13a to 2.13d) and after 4sec. the now, random motion was almost non-existant. Gravitational collapse of the mixing region produced reverse flow just above the body (Fig. 2.13e to 2.13h) and, the upper part of the mixing region still moved in the forward direction. Further from the ramp, flow within the mixing region was all in the forward direction.

The body thickness behind the prominent front of the reverse flow was greater than in front of it (Fig. 3.13h). This would suggest that the front of the reverse flow was essentially a moving hydraulic jump (surge) that then advanced into the tail of the forward flow.

The height of the bulge measured at the ramp, can be assessed as a non-dimensionalized value of D/H, where D is the height of the bulge (cm), and then plotted against the density ratio ρ_1/ρ_2 of the two fluids (Fig. 2.14):

Plate 2.3

(overleaf)

Vertical views of the fluid motion that occurs as a forward-moving, surge style density current is incident upon a 20° ramp. Time (t) is in seconds after Plate 3.3a.

(a) t = 0s. The head of the forward current begins to travel up the ramp.

(b) t = 1s. The head of the forward density current progressively thins as it moves up the ramp. The body of the flow has thickened slightly at the foot of the ramp.

(c) t = 2s. A distinct bulge of fluid is now visible at the foot of the ramp. The head of the forward-moving current is now approximately half the thickness that it was originally.

(d) t = 3s. The bulge of fluid propagates away from the ramp. As it propagates, it cuts under the mixing zone of the remaining forward-moving density current.

a

b

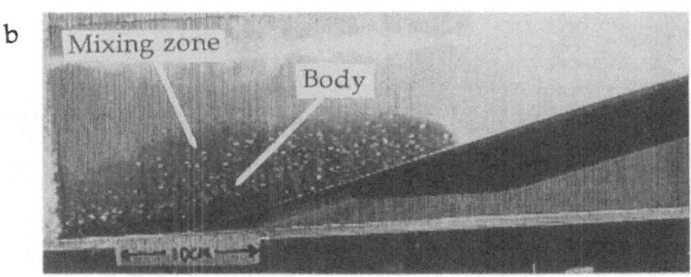

c

d

Plate 2.3

10cm

83

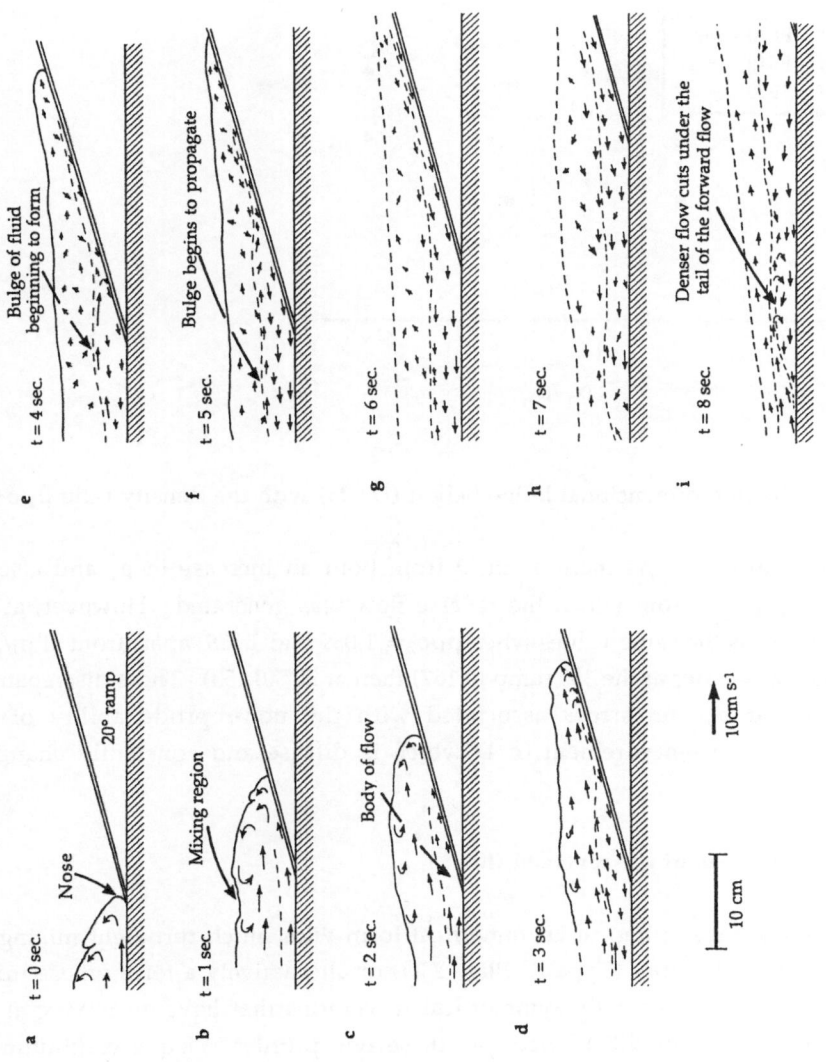

Fig. 2.13

Illustration of the internal fluid motion that occurs when a density current is incident upon a 20° ramp in the rectangular flume tank. The arrows indicate Pliolite particle displacements.

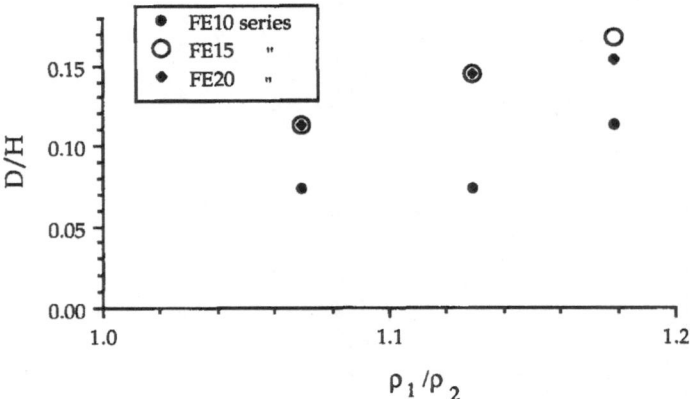

Fig. 2.14

Variation of the non-dimensional bulge height (D/ H) with the density ratio ρ_1/ρ_2.

There is some indication for the increase in D from both an increase in ρ_1 and also an increase in the ramp angle from which the reverse flow was generated. However, at the 15° and 20° ramps, D has the same values when $\rho_1/\rho_2 = 1.069$ and 1.129, apart from at $\rho_1/\rho_2 = 1.179$ where D/ H was greater at the 15° ramp (0.167) than at 20° (0.153). These discrepancies can be accounted for by the errors associated with the non-reproducability of the experiment and with the measurement of D which is diffuse and constantly changing (Appendix 2A).

2.7.4 Variation in the form of the reversed flow

The reverse flow may have either an asymmetrical form with much turbulent mixing on the downstream side of the front (Type A, Plate 2.2a) or alternatively a form whose upper surface resembles a group of smooth, symmetrical waveforms that have no mixing at the upper interface (Type C, Plate 2.2c). Between these two profiles exists a continuum of intermediate profiles that have varying degrees of symmetry and mixing (Type B, Plate 2.2b). The variation of the form of the reverse flow, its internal nature and the factors which govern which type of reverse flows are generated are examined in the following sections. The mean thickness of the reverse flow is defined as d_1 (Fig. 2.15) which may be non-dimensionalized as the ratio d_1/d_0, where d_0 is the flow depth of the tail of the forward flow immediately prior to the arrival of the reverse flow.

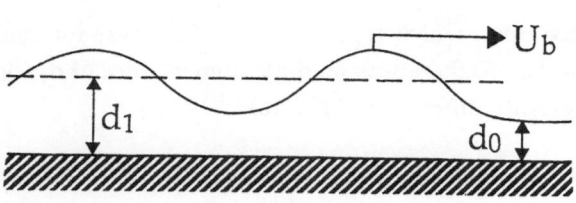

Fig. 2.15
Definition of the mean flow depth of the reverse flow (d_1) and the thickness of the flow prior to the arrival of the reverse flow (d_0).

Comparison of d_1/d_0 for the S10 to S20, FE10 to FE20 and CD10 series of experiments with the density ratio ρ_1/ρ_2, specifying which type of reverse flow formed, showed that: (i) there

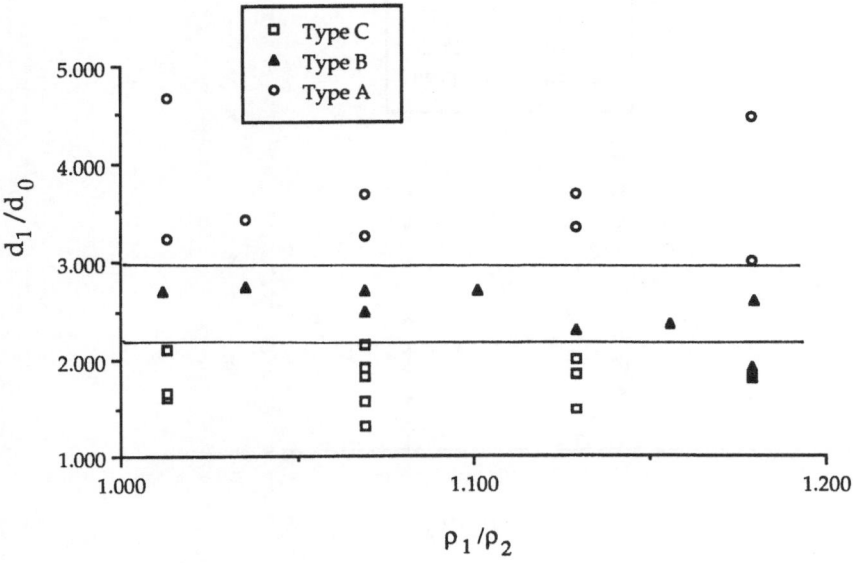

Fig. 2. 16
Variation of the fractional revers flow thickness (d_1/d_0) with the density ratio ρ_1/ρ_2 showing the dependence of the type of reverse flow formed and the non-dimensional reverse flow depth.

was a strong dependence upon which type of reverse flow was formed with d_1/d_0, (ii) d_1 was always greater than d_0, (iii) there was no direct dependence of d_1/d_0, or type of reverse flow formed, upon the density difference (Fig. 2.16).

Which type of reverse flow is formed is also controlled, to some extent, by the ramp angle upon which the original current had been incident. For example, it can be seen on Figure 2.17, that for a particular density d_1/d_0 increased as the ramp was steepened. It is also apparent that the results for the S10 to S20 series of experiments are greater than for the FE10 to FE20 series of experiments. The trends are affected by scatter, which is worse in the S10 to S20 series of experiments, due to the difficulty of accurate measurement of a diffuse boundary. The large measurement error could also account for the trend discrepancies. These discrepancies also occurred in the FE10 to FE20 series of experiments where in addition, the ramp angle apparently had a smaller effect in determining the value of d_1/d_0.(Fig. 2.17).

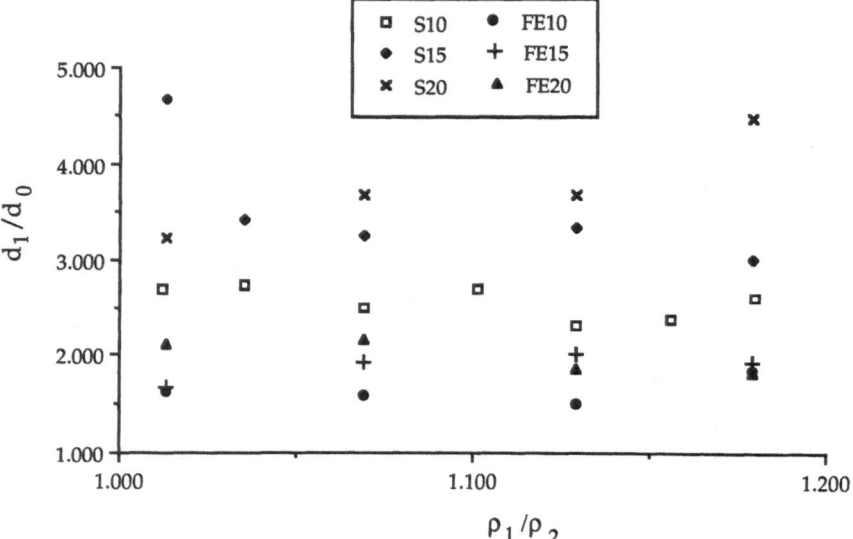

Fig. 2.17

Variation of fractional reverse flow thickness (d_1/d_0) with the density ratio (ρ_1/ρ_2) for the S10 to S20 and FE10 to FE20 series of experiments.

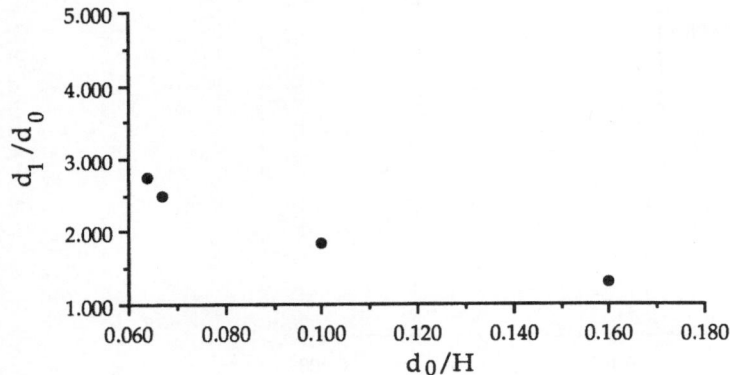

Fig. 2.18
Variation of fractional flow thickness (d_1/d_0) depth with fractional flow depth for the CD10 series of experiments.

The mean flow depth was also dependent upon the fractional depth of the flow, d_0/H (Fig. 2.18), with d_1/d_0 decreasing as the d_0/H increased, ie. at higher fractional flow depths the reverse flow was more likely to have the profile of a train of waves rather than a turbulent, unsteady current.

The 'amplitude' of the waveforms can be described as (a_t-d_2), where a_t is the maximum height of the wave from the base of the tank and d_2 is the flow depth **after** the wave (Fig. 2.19 a and b).

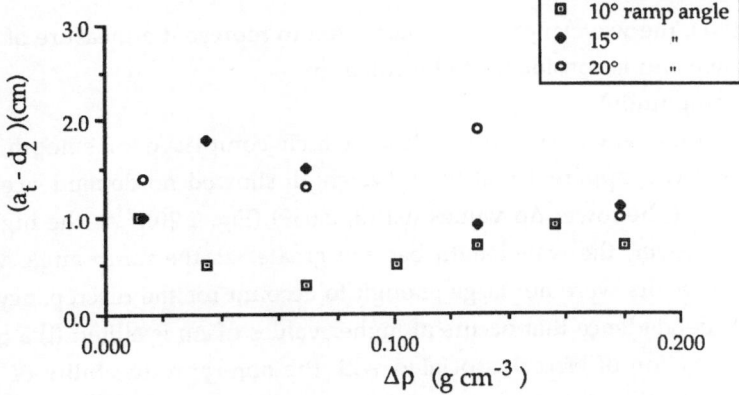

Fig. 2.19 a

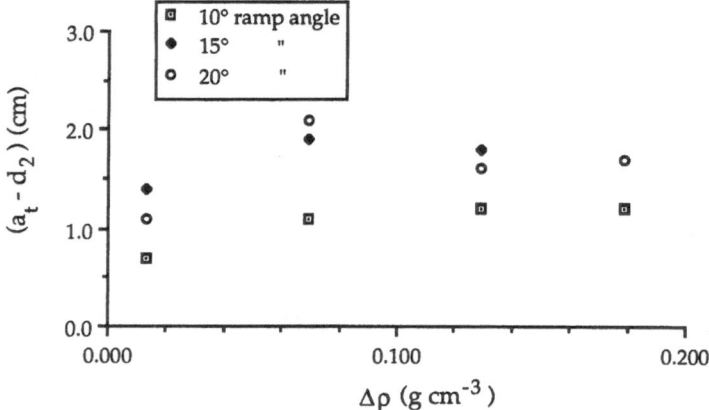

Fig. 2.19 b

Fig. 2.19 a and b
(a) Variation of amplitude (a_t -d_2) with density for the S10 to S20 series of experiments.
(b) Variation of (a_t -d_2) with density for the FE10 to FE20 series of experiments.

The amplitude was largely independent of $\Delta\rho$ and the fractional depth of the flow (d_1/H), but showed some dependence on the ramp angle, increasing between angles of 10° and 15° (Figs. 2.19 a and b). However, in S10/1.012 the amplitude had a constant value of 1.0 at all ramp angles. Despite the scatter, there was no obvious trend between 15° and 20°, in either the the S10 to S20 or FE10 to FE20 series' of experiments.

After Koop and Butler (1981), the wavelength, L is considered to represent a measure of the horizontal extent of the wave and is considered to be equal to:

$$L = 0.5 \text{ (area/amplitude)}$$

Considering only the the first waveforms from flows which comprised of smooth, or nearly smooth, waveforms it was apparent that the wavelength showed no dependence on either $\Delta\rho$ or the ramp angle at the lower $\Delta\rho$ values (0.013, 0.069) (Fig. 2.20). At the higher $\Delta\rho$ values (0.129, 0.179), however, the wavelength became greater as the ramp angle was increased. As measurement errors were not large enough to account for the discrepancy, it must be supposed that the dependence that occurs at higher values of $\Delta\rho$ is either: (i) a true feature or, (ii) that it is a function of errors associated with the non-reproducability of the experiments. Unfortunately, the data presented does not allow a more significant conclusion to be drawn.

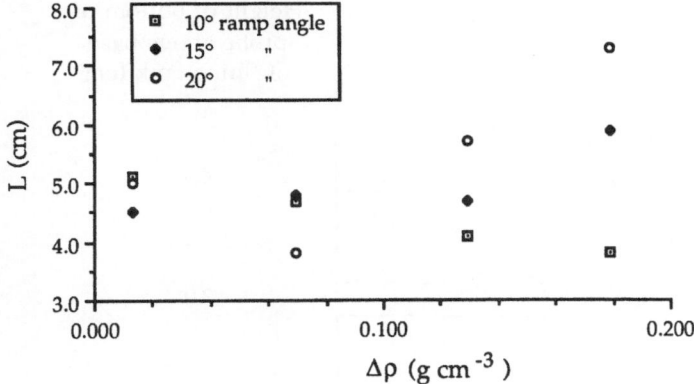

Fig. 2.20

Variation of the wavelength (L) of the reverse flow waveforms with the density difference ($\Delta\rho$) for the FE10 to FE20 series of experiments.

As previously stated, many of the reverse flow exhibited characteristics in common with forward flow density currents:

(i) Where billows occurred (eg. S15/1.068), they were a continuous feature across the crest of the waveform.

(ii) In several of the reverse flows, lobes and clefts were observed. The flows in which they occurred were very similar in form to a forward flow density current. They occurred only at the edges of the 'waveform'; the very centre of the front of advance of the 'waveform' remaining smooth.

(iii) The presence of a nose at the front of the reverse flow was only occasionally observed (eg. S20/1.128). This could be a result of the visualisation technique being inadequate, as much as the absence of the nose.

2.7.5 Velocity profile of a type C reverse flow

Figure 2.21 illustrates the variation in velocity as the reverse flow, which in this instance consists of a series of smooth, symmetrical waveforms, passed a stationary point in the centre of the flume tank. As with the passage of the head of the current (Fig. 2.3) the greatest velocities occurred in the basal centimetre of the flow, with a rapid decrease in the magnitude of the velocities vertically through the profile. In contrast to the passage of the density current, the propagation of the symmetrical waveforms resulted in a symmetrical velocity profile as each individual wave passed the probe. Between the waves, the velocity was again constant decreasing to almost zero by 3cm above the base of the tank. This would suggest that each wave was an independent feature. The variation in maximum velocity that occurs vertically is a result of experimental non-reproducability.

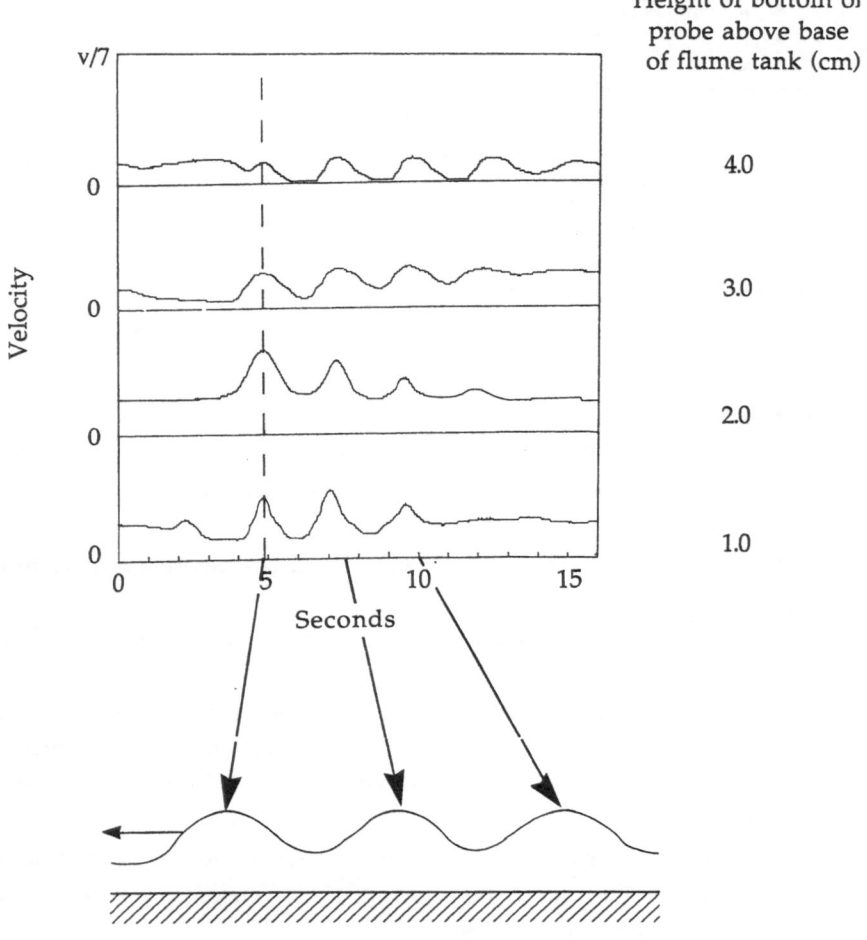

Fig. 2.21

Relative velocity profiles as the reverse flow passes a fixed point in the centre of the rectangular flume tank. The maximum measured velocity of the wave is 6.3cm s⁻¹. This will be similar to the maximum velocity in the profile measured at 1.0cm above the base of the flume tank. The vertical velocity scale (v) corresponds to that in Fig. 2.3.

2.7.6 Variation of crestal velocities of the reverse flow

The velocity of the front of the reverse flow in the S10 to S20 series of experiments, and of the first wave of the reflected flow of the FE10 to FE20 series increased as the density of the saline solution became greater (Figs. 2.22 a and b).

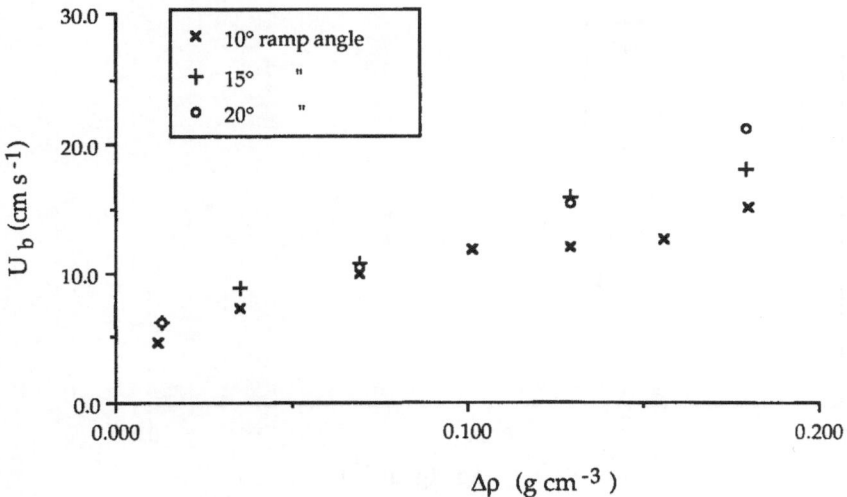

Fig. 2.22 a
Variation of velocity of the front (U_b) of the reverse flow with $\Delta\rho$ for the S10 to S20 series of experiments.

In both groups of experiments the velocity increased as an approximately linear function of the density of the saline solution (Figs. 2.22 a and b). There was also a slight increase in the velocity, for a particular density, as the ramp angle was steepened. This relationship was more apparent at the higher values of $\Delta\rho$ where scatter due to the non-reproducability of the experiments was a smaller percentage of the total velocity. In addition, the ramp angle would appear to have had a greater effect between the 10° and 15° ramp angles. The volume of saline fluid from the body that accumulated at the ramp was, similarly, significantly greater between the 10° and 15° ramp angle than between the 15° and 20° ramps (Fig. 2.14). It would therefore, be logical to suggest that the velocity of the reverse flow was dependent upon the volume of fluid that accumulated in the elevated body that formed at the ramp, as greater volumes will have a greater potential energy. The velocities between the S10 to S20 and the FE10 to FE20 series' of experiments were very similar for equivalent values of $\Delta\rho$ although the forms of the reverse flow represent both of the end-member profiles (Fig. 2.22 a and b). It is apparent, therefore, that the absolute velocity of the reverse flow was not the most important factor controlling the form of the reverse flow. The Reynolds numbers of the reverse flow varied from 80 at the lowest densities used (1.012 g cm^{-3}) to 440 at the highest densities (1.178g cm^{-3}) Thus, viscous forces will be an additional factor affecting the velocities of the reverse flows.

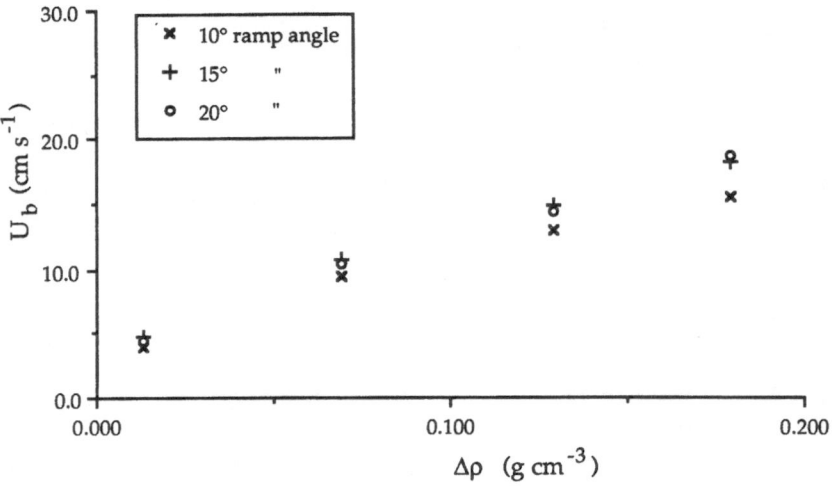

Fig. 2.22 b
Variation of velocity of the first waveform (U_b) of the reverse flow with $\Delta\rho$ for the FE10 to FE20 series of experiments.

The variation of U_b of the reverse flow (in this instance, a train of smooth waves) with d_1 is illustrated in Fig. 2.23. Overall, the velocity of the first wave of the reverse flow increased as the mean flow depth (d_1) became greater. However, either the velocity when d_1 was 1.6 or 1.7 are not in agreement with this trend, with the respective velocities being too great or too small. Such a large discrepancy could be a result of the non-reproducibility of the experiments. If this is the case then any suggestion of a trend could also be artificial being generated by the aforementioned effects. However, the value of d_1 of 1.6 is associated with the greatest breadth : depth ratio in the flume tank (2.7:1) and therefore, according to Barr (1963) the smallest edge effects. Thus the high velocity when d_1 is 1.6cm could, at least in part, be due to smaller edge effects.

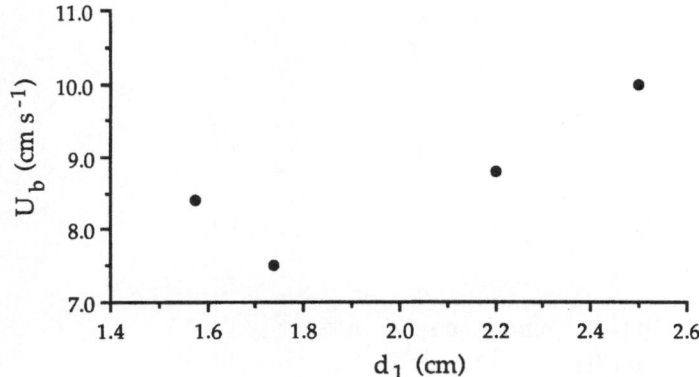

Fig. 2.23
Variation of velocity (U_b) with the mean thickness (d_1) of the reverse flow for the
CD10 series of experiments.

For the S10 to S20 series of experiments, U_b was approximately 70 to 100% of U_h and largely
independent of $\Delta\rho/\rho_1$, although at higher values of $\Delta\rho/\rho_1$ (0.129, 0.179) and at steeper ramp
angles, U_b became increasingly similar to U_h (Fig. 2.24).

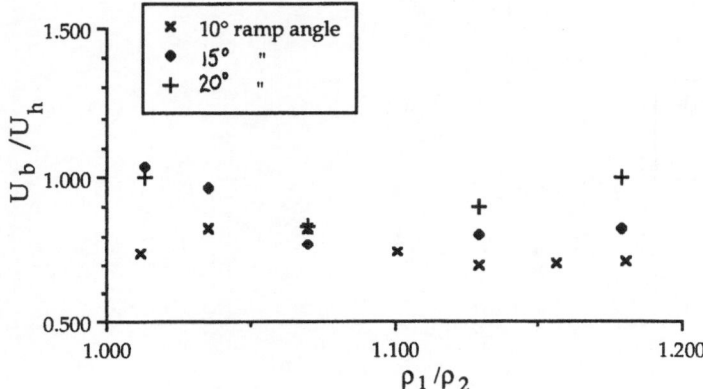

Fig. 2.24
Variation of the fractional velocity U_b/U_h with the density ratio ρ_1/ρ_1 for the S10 to
S20 series of experiments.

The variation in the velocity of the reverse flow in the CD10 series of experiments varied
between 74 and 100% of the velocity of the head of the forward density current (Fig. 2.25).

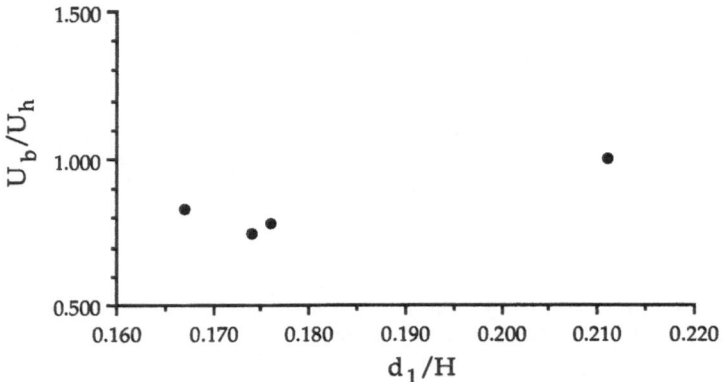

Fig. 2.25

Variation the fractional velocity (U_b/U_h) with fractional flow depth (d_1/H) for the CD10 series of experiments.

Similar considerations for the FE10 to FE20 series of experiments were not in agreement with those of the S10 to S20 series of experiments. In these, U_b/U_h, although still independent of $\Delta\rho/\rho_1$, indicated that the velocity of the reverse flow was equal to, or greater than the velocity of the head of the forward current, up to a maximum value of U_b of approximately $1.3U_h$ (Fig. 2.26).

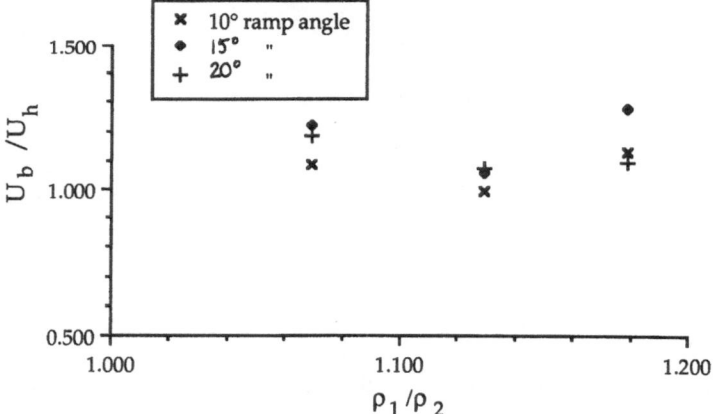

Fig. 2.26

Variation of fractional velocity (U_b/U_h) with the density ratio ρ_1/ρ_2 for the FE10 to FE20 series of experiments.

In most of the experiments conducted, by the time the reverse flow had travelled as far as the end wall of the lock, it had the profile of a group of waves. The velocity of the wave reflected from the wall was 70 to 90% of the velocity of the equivalent wave prior to interaction with the vertical end wall of the tank. As the waves travelled back towards the ramp, they lost a further 15 to 44% of their velocity. As no mixing due to shear instabilities occurred, the loss of velocity must have resulted from frictional damping at the flume edges, viscosity and also from internal shear with the opposing flow.

2.7.7 The numbers of waves associated with a weak bore

Usually four well defined waves were produced during the generation of an undular bore in the rectangular tank. Behind the well-defined waves existed a further group three or four waves that become progressively smaller in amplitude and were often difficult to identify. The numbers of waves produced were similar between experiments and there was no trend with, for example, density, ramp angle or depth of fluid. Within the bounds of these experiments, differences in the wave numbers generated were most probably due to the non-reproduciblity of the experiments.

2.7.8 Internal flow characteristics of the reversed flow

Prior to the arrival of the reverse flow, in the experiments S15/1.012 to S15/1.178, the velocity from Pliolite particle paths within the tail, increased from zero (Fig. 2.27 a) to approximately 5.4cm s^{-1} (Fig. 2.27 c and d). The motion was bed-parallel.

As the reverse flow impinged upon the residual tail of the forward flow, the form of the reverse flow was affected by the density difference. At the lowest density (Fig. 2.28 a), the flow had the form of a diffuse asymmetrical wave. Motion within this wave was bed-parallel, attaining a maximum velocity of 5.9cm s^{-1} in the central portion of the wave which decreased to 1.6cm s^{-1} near the base. However, at higher densities, the flow resembled a density current with an asymmetrical form, billows and a finer-scale of mixing. The billows were most clearly defined when $\rho_1 = 1.068$g cm^{-3} (Fig. 2.28 b). At higher values of ρ_1 (Figs. 2.28 c and d), a finer scale of mixing dominated over any larger scale structures and entrainment of fluid from the mixing

Figs. 2. 27 to 2.29
Internal motion immediately before (Fig. 2.27a to d), during (Fig. 2.28a to d) and after (Fig. 2.29 a to d) the arrival of the reverse flow front for the S15 series of experiments. The arrows indicate Pliolite particle displacements over 0.32 seconds in a Eulerian reference frame. The salt solution - fresh water interface is indicated by a solid line at t = 0s and a dashed line at t = 0.32s.

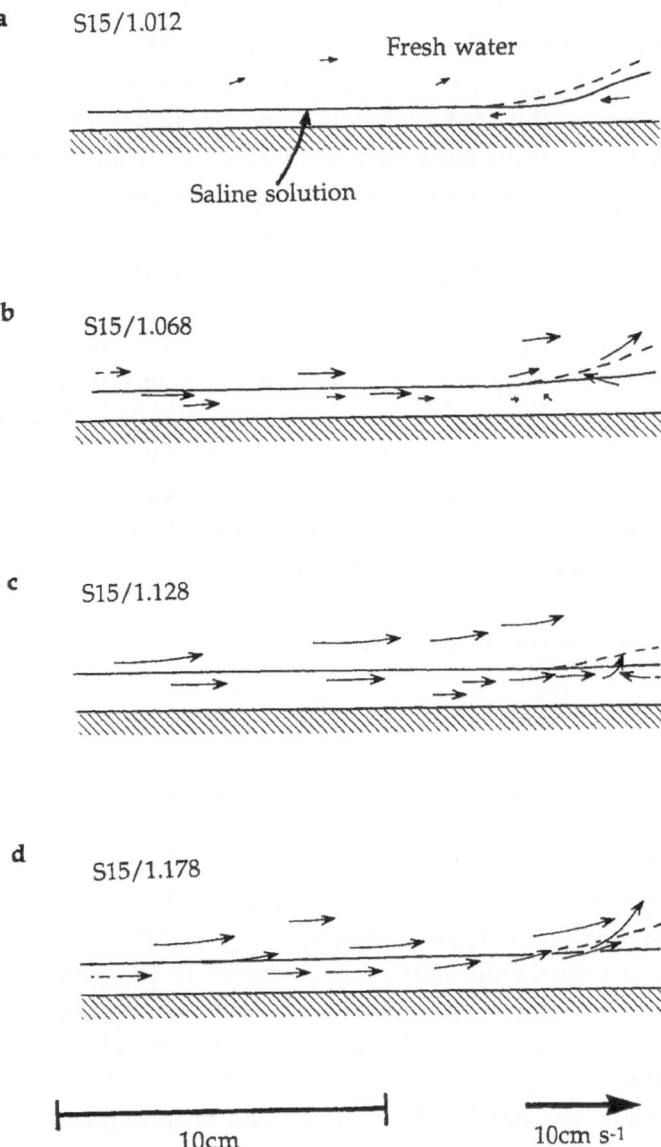

a S15/1.012

Fresh water

Saline solution

b S15/1.068

c S15/1.128

d S15/1.178

10cm

10cm s⁻¹

Fig. 2.27 a to d
Immediately before the arrival of the reverse flow front.

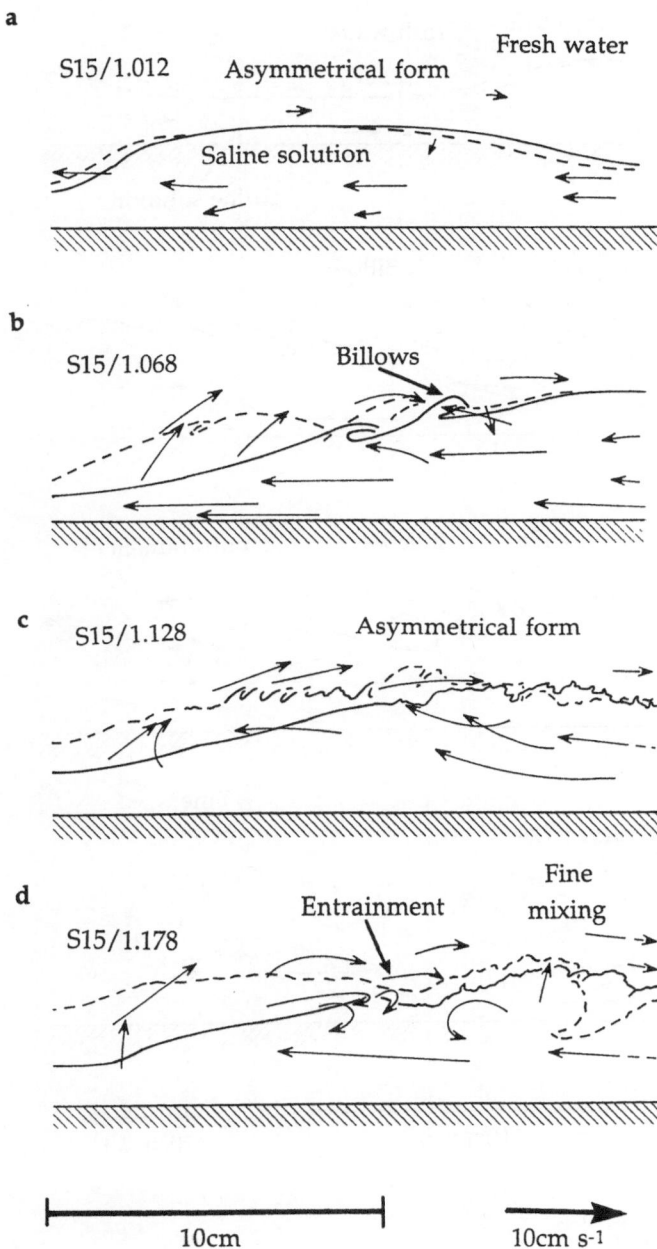

a

S15/1.012   Asymmetrical form   Fresh water

Saline solution

b

S15/1.068   Billows

c   S15/1.128   Asymmetrical form

d   Entrainment   Fine mixing

S15/1.178

|— 10cm —| 10cm s⁻¹

Fig. 2.28 a to d
The arrival of the reverse flow front.

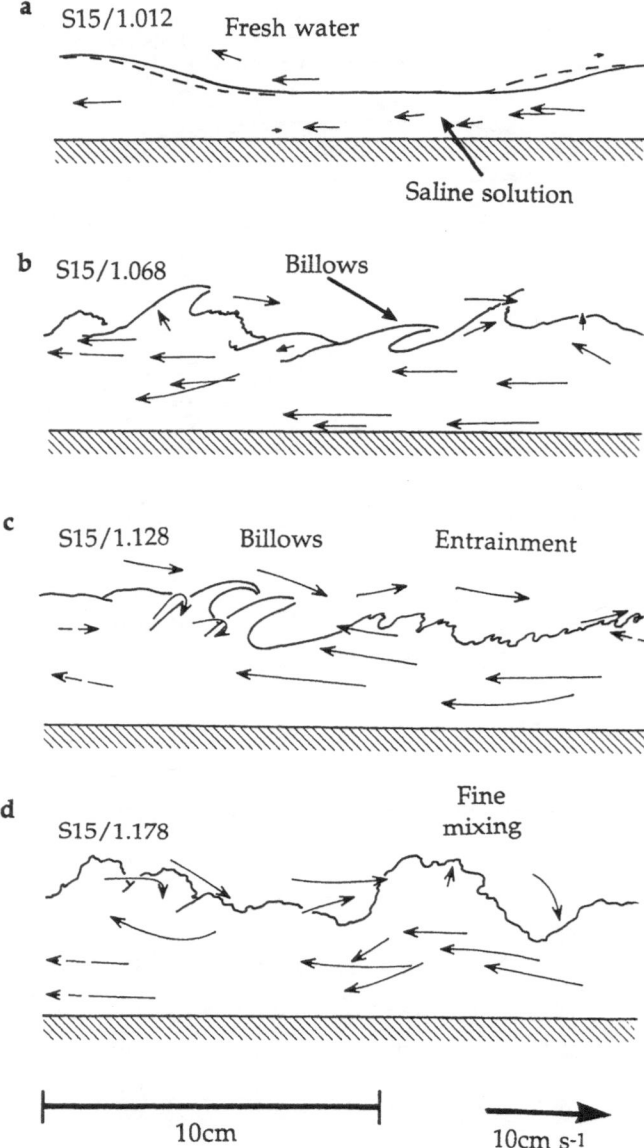

Fig. 2.29 a to d
Behind the front of the reverse flow.

region into the reverse flow occurred more vigorously. Overall, velocities within the reverse flow increased as ρ_1 became greater, although between S15/1.128 and S15/1.178 the maximum velocities were similar, at approximately 17.2cm s^{-1}, in the central portion of the wave. Apart from the particles that had become entrained into the reverse flow, fluid motion within the mixing region remained in the forward direction and was pushed up and over the reverse flow as it intruded underneath.

Between the first and second waveform of the reverse flow of S15/1.012 (Fig. 2.29 a) the motion was bed-parallel with a maximum of 3.2cm s^{-1} in the direction of the reverse flow. In the denser flows (Figs. 2.29 b, c and d) individual waveforms were not observed and instead two 'humps' of fluid occurred. Although remaining approximately parallel to the bed, the fluid motion behind the first 'mass' was increasingly random, as ρ_1 became greater. The maximum velocity of 11.9cm s^{-1} occurred within S15/1.128 and S15/1.178.

The motion of individual particles within the reverse flow in a Lagrangian reference frame (frame of reference moves with the wave) are illustrated in Figures 2.30 a, b, c and d. At low values of ρ_1 (1.012, 1.068g cm^{-3}) observation of individual particles for the time it took for the front of the reverse flow to pass the field of view were possible. On comparison of Figure 2.30a with Figure 2.28a it can be seen that, although net motion was in the reverse direction, particles near the base and upper surface of the body of the flow moved slower than the crestal velocity (U_b) of the waveform. As a result, particles became disentrained as the waveform propagated through the flow. Conversley, particles within the central portion of the flow were transported at approximately the same velocity as the wave for some time although ultimately, they became disentrained. This means that long distance transport is possible. At higher values of ρ_1, similar trends were visible but poor particle visualisation, due to both the velocity of travel and turbulent mixing complicated the analysis.

The reverse flow in the FE15 series of experiments consisted of a train of smooth, symmetrical waves with no mixing at the upper surface (Figs. 2.31 to 2.34). Velocities within the tail prior to the arrival of the reverse flow were zero in FE15/1.012 and attained a maximum of 0.6cm s^{-1} (although this was in the reverse direction) in FE15/1.178 (Fig. 2.31d). Prior to the arrival of the reverse current there was a noticable drift of particles in

Figs. 2. 31 to 2.33
Internal motion immediately before (Fig. 2.31a to d), during (Fig. 2.32a to d) and after (Fig. 2.33 a to d) the arrival of the reverse flow front for the FE15 series of experiments. The arrows indicate Pliolite particle displacements over 0.32 seconds in a Eulerian reference frame. The salt solution - fresh water interface is indicated by a solid line at t = 0s and a dashed line at t = 0.32s.

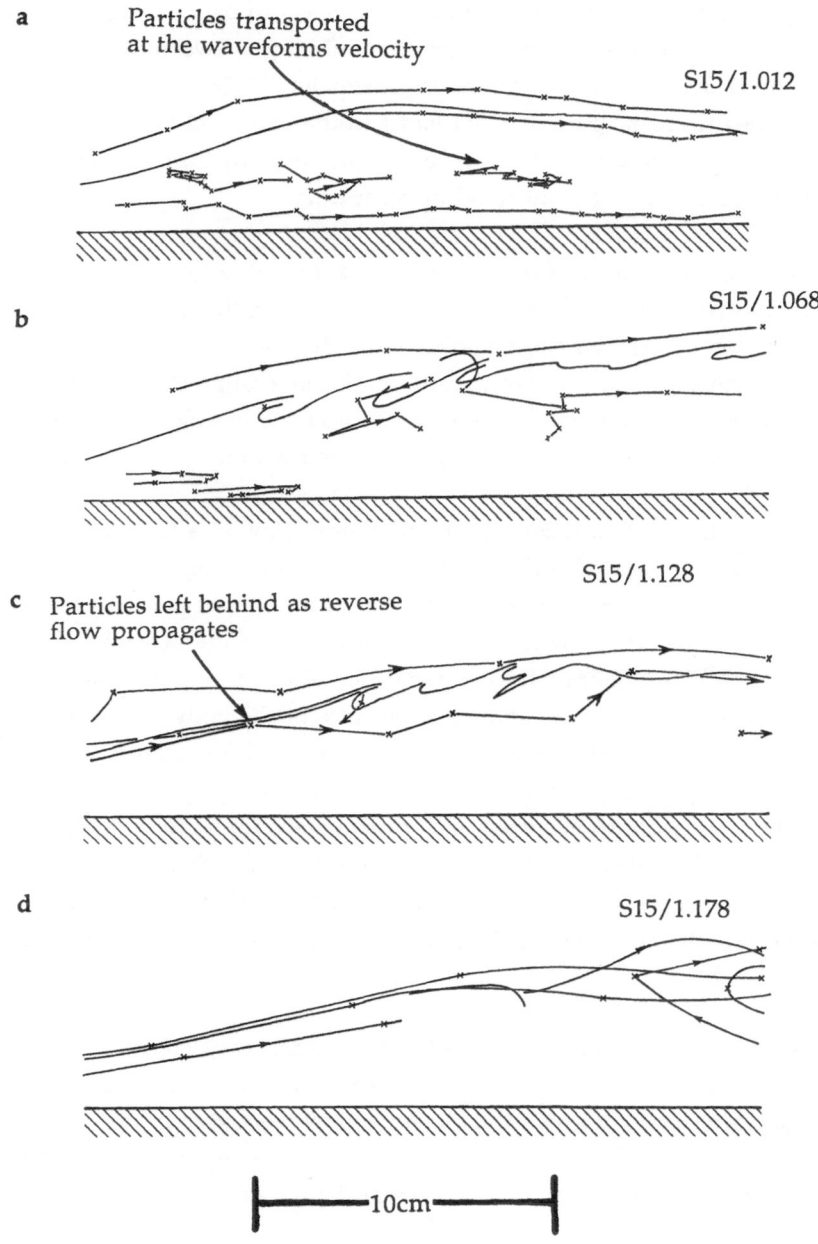

a   Particles transported
    at the waveforms velocity   S15/1.012

b   S15/1.068

                                                  S15/1.128

c   Particles left behind as reverse
    flow propagates

d   S15/1.178

|——————10cm——————|

Fig. 2.30 a to d

Pliolite particle movement in a Lagrangian reference frame (ie. the
particle motion is analysed relative to a stationary reverse flow front) for
the S15 series of experiments. The Pliolite particles are marked at 0.24s
intervals (x)

a FE15/1.012 - No particle movement.

b FE15/1.068

c FE15/1.128

d FE15/1.178

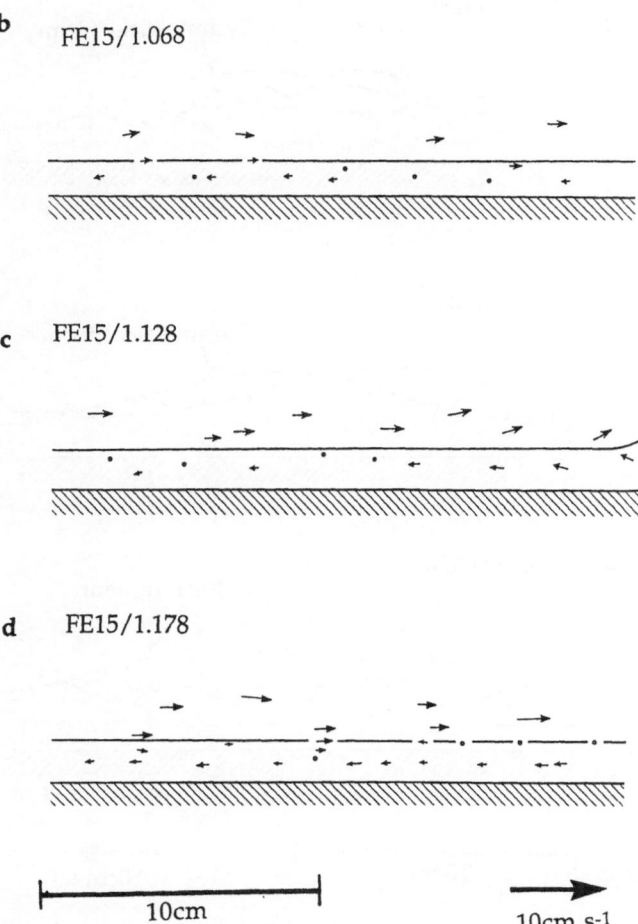

|———————————————| →
        10cm   10cm s⁻¹

Fig. 2.31 a to d
Immediately before the arrival of the reverse flow front.

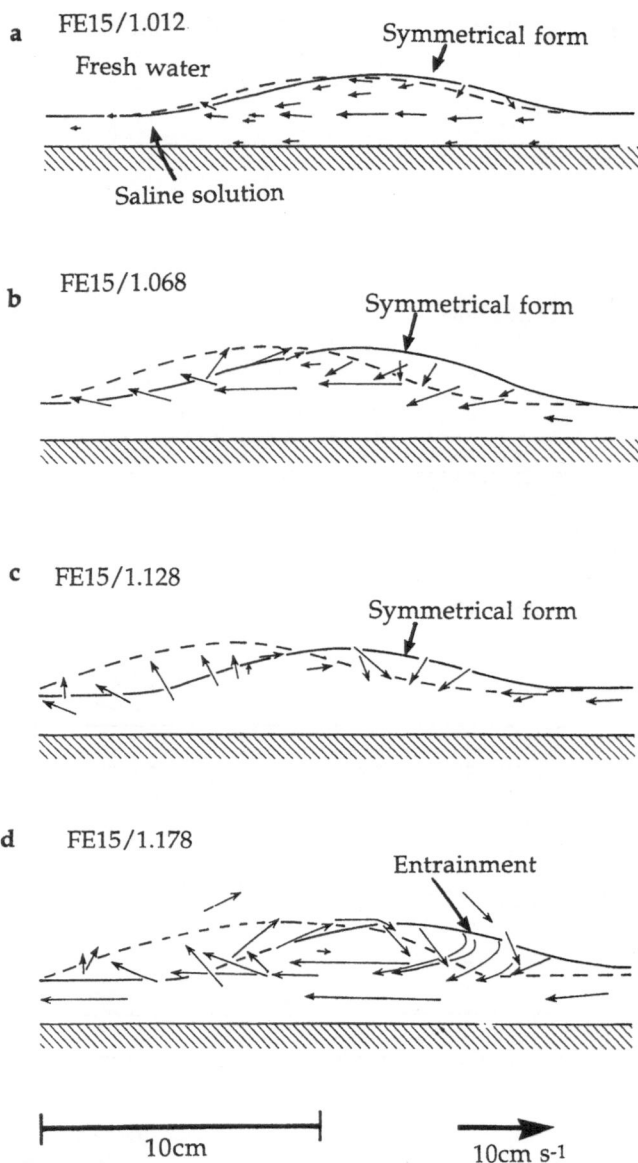

a FE15/1.012

Fresh water

Symmetrical form

Saline solution

b FE15/1.068

Symmetrical form

c FE15/1.128

Symmetrical form

d FE15/1.178

Entrainment

10cm

10cm s⁻¹

Fig. 2.32 a to d
The arrival of the reverse flow front.

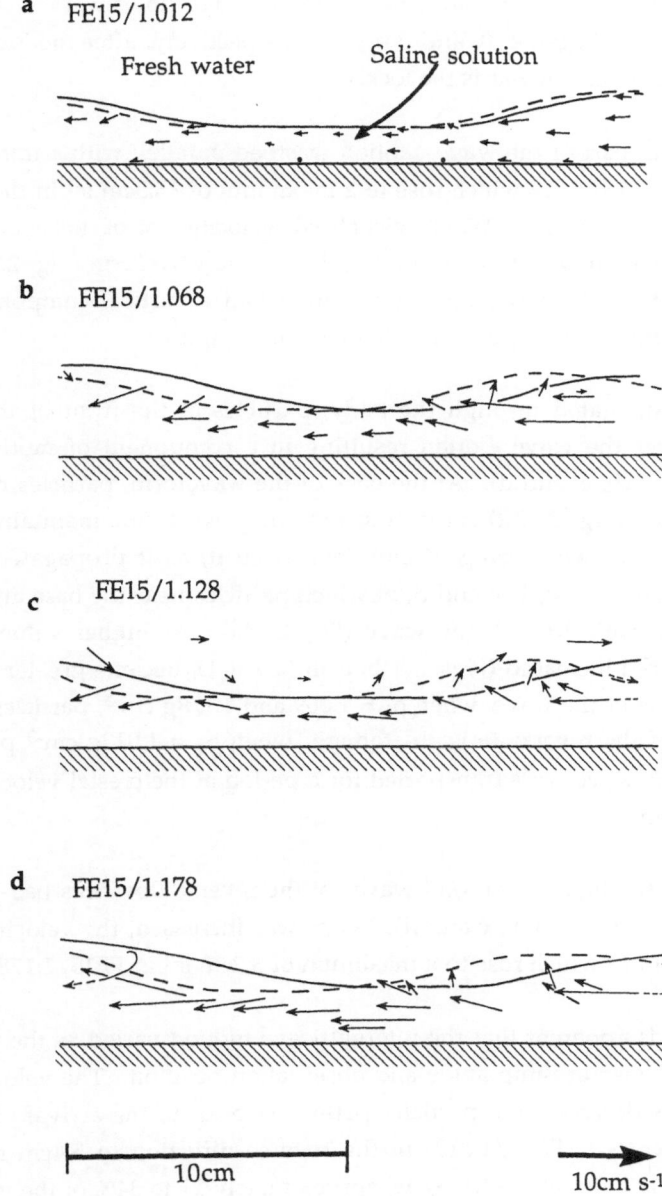

Fig. 2.33 a to d
Behind the front of the reverse flow.

the reverse direction within the lower part of the tail. For flows FE15/1.068, FE15/1.128 and FE15/1.178 this drift began 4, 0, and 0 seconds respectively, after the 'bulge' of fluid at the ramp began to propagate towards the lock.

Within the central part of the wave, motion was bed-parallel, with a minimum of 2.0cm s^{-1} near the base in FE15/1.012 which rose to a maximum of 15.3cm s^{-1} in the central portion of the first wave of FE15/1.178. Where visualised, velocities of particles on and near the bed were slightly less than within the central region of the waveform (Fig. 2.32a). In the upper 0.5 to 1.0cm of the waveform, the fluid motion had a vertical component. This vertical component became more important at higher ramp angles.

As the wave propagated through the body, particles at the front of the waveform were pushed back over the wave - often resulting in a component of motion in the forward direction (Figs. 2.32b, c, and d). At the back of the waveform, particles moved towards the base of the wave (Fig. 2.32d) indicating that they were not maintaining their position within the wave but were being disentrained from it, as it propagated. Such motion is illustrated in Figures 2.34a, b, c and d, in which particles near the base and upper surface of the flow moved back through the wave (Fig. 2.34a). At higher values of ρ_1 the rate at which this occurred increased (Figs. 2.34b, c and d) as U_b became greater. However, within the central portion of the wave, when ρ_1 = 1.069 and 1.178g cm^{-3}, particles were transported at the velocity of the reverse flow. Although, when ρ_1 = 1.012g cm^{-3} particles within the central part of the wave were transported for a period at the crestal velocity but after a time were disentrained.

Motion between the first and second waves of the reverse flow was bed-parallel and in the reverse direction (Figs. 2.33, b, c and d). As ρ_1 was increased, the velocity had a minimum of zero in FE15/1.012 which rose to a maximum of 3.2cm s^{-1} in FE15/1.178.

In conclusion, it is apparent that the internal wave motion varied as the initial density was increased, for a constant ramp angle and observation position. The velocity within the tail of the current, as derived from particles paths just prior to the arrival of the reverse flow, varied from 1.0cm s^{-1} (FE15/1.012) in the reverse direction to 5.4cm s^{-1} in the forward direction (S15/1.128 and FE15/1.178), ie. approximately 21 to 34% of the velocity of the crest of the first wave. The velocity of the tail therefore, cannot always be dismissed as negligable, as suggested by Pantin and Leeder (1987) in their experiments, and must be thought of as another factor affecting the nature of the reverse flow. As regards sediment transport, these studies of particle motion indicated that the mass transport of sediment, for a considerable distance, could occur within the central portion of the reverse flow. In addition, bed-load may become suspended as the wave passes but will settle out again once the waveform or 'hump' of fluid has passed.

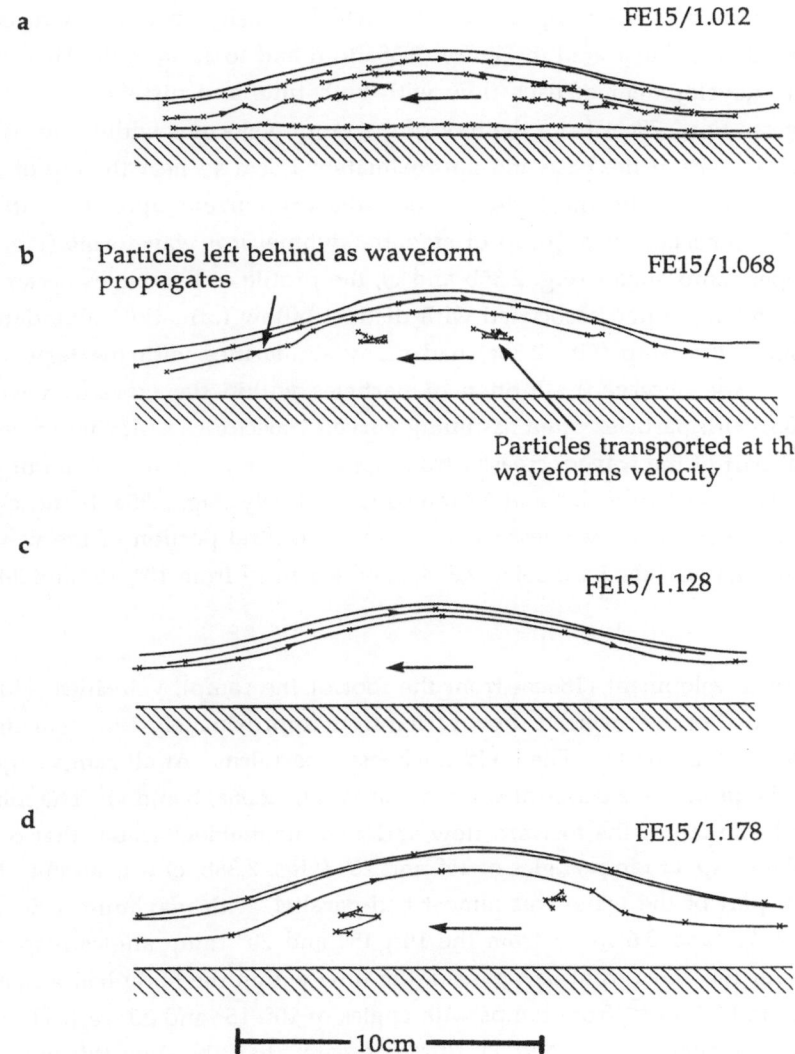

a   FE15/1.012

b   Particles left behind as waveform
     propagates   FE15/1.068

Particles transported at the
waveforms velocity

c   FE15/1.128

d   FE15/1.178

⊢——— 10cm ———⊣

Fig. 2.34

Pliolite particle movement in a Lagrangian reference frame (ie. the
particle motion is analysed relative to a stationary waveform) for the
FE15 series of experiments. The Pliolite particles are marked (x) at 0.24s
intervals.

The influence of increasing the ramp angle upon which density currents, with constant initial ρ_1, were incident is illustrated in Figures 2.35a to d and to 2.36a to d. These Figures also illustrate the development of each flow with both time and distance. Prior to the arrival of the reverse flow, 79cm from the foot of the ramp, velocities within the tail of the flow varied between zero at the base and approximately 1.5cm s^{-1} near the top of the tail. The reverse flow generated by the incidence of a density current upon a ramp whose inclination was 10° consisted of a group of smooth, symmetrical waveforms (Fig. 2.35a). However, at steeper ramp angles (Fig. 2.35b and c), the profile of the waves became more asymmetrical and mixing, often associated with distinct billow formation, abundant. The reverse flow from a 20° ramp (Fig. 2.35c) had many similarities with the forward flow (although no nose was observed). Motion of particles within the waveform was sub-parallel to the bed, with particles often assuming curved trajectories. Maximum velocities within the central part of the wave increased from approximately 7.0cm s^{-1} at a ramp angle of 10°, to 9.8 and 11.6cm s^{-1} from 15° and 20° ramps respectively (Fig. 2.35a, b and c). Near the base of the flow, velocities were less than within the central portion of the waveform, with measured maxima near the base being 2.8, 4.3 and 8.3cm s^{-1} from 10°, 15° and 20° ramp angles respectively.

At a later stage of development (186cm from the foot of the ramp), velocities within the tail, prior to the arrival of the reversed flow, varied from approximately zero near the base, to 1.2cm s^{-1}· higher within the tail. The body thickness was 1.0cm. At all ramp angles, the reverse flow had the profile of a series of smooth waves (Fig. 2.36a, b and c). The mixing of the flow with both the tail of the forward flow and also the ambient water, that occurred 107cm closer to the ramp, at ramp angles of 15° and 20° (Figs. 2.36b, c) was absent. Motion within the central part of the wave was almost bed-parallel, with maximum velocities of approximately 6.7, 7.7 and 8.6cm s^{-1} from the 10°, 15° and 20° ramp angles respectively. Fifteen seconds after the wave front had passed, the flow within the body had a maximum velocity of 3.1, 4.5 and 2.2cm s^{-1} from ramps with angles of 10°, 15° and 20° respectively. At this stage, the flow thickness was 1.6cm (1.6 times greater than the flow thickness in the residual tail).

Apart from the conclusion that increasing the ramp angle affects the form of the reverse flow observed at a stationary point, it is also apparent that, an asymmetrical reverse flow, with abundant mixing will, in time, become a train of smooth, symmetrical waves with little or no mixing.

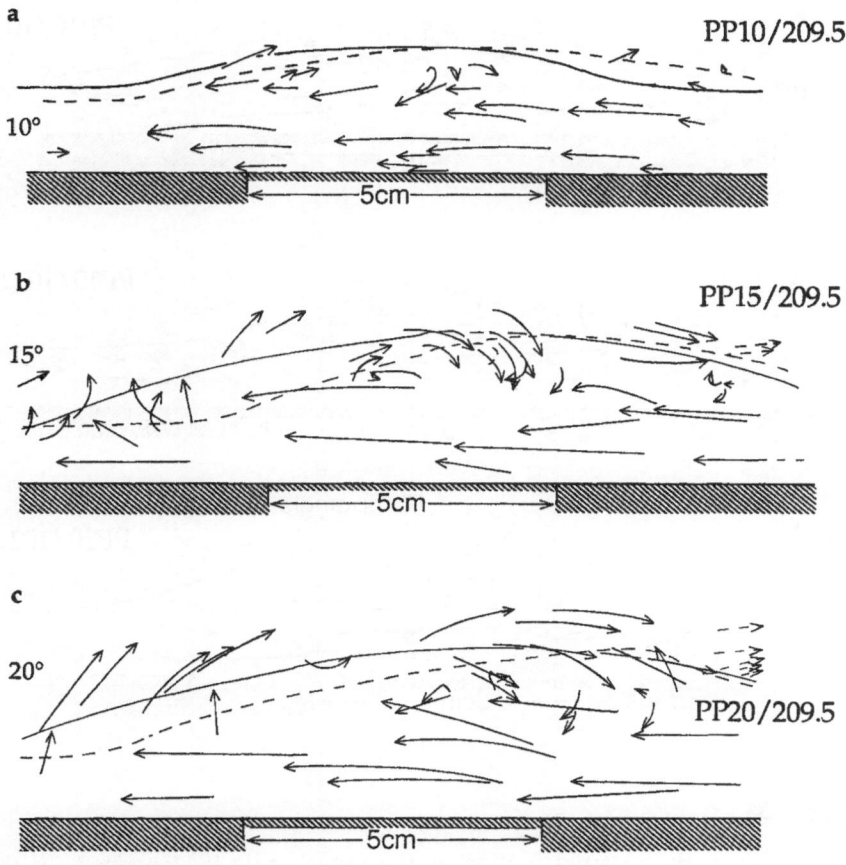

Fig. 2.35

Variation of the internal motion as the ramp angle is steepened for the PP10 to PP20 series of experiments. The observation point is 209.5cm from the end wall of the lock.

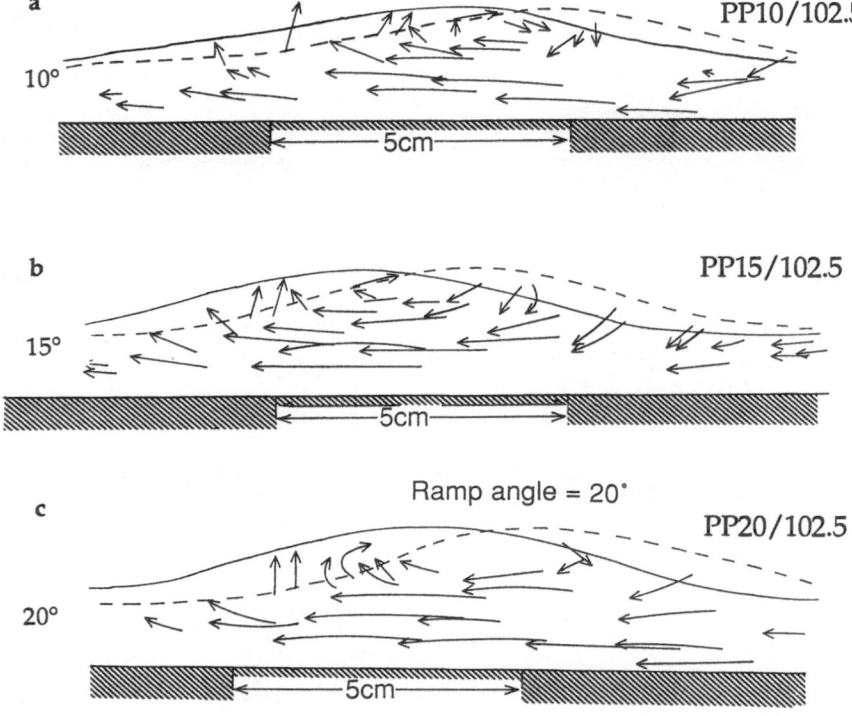

Fig. 2.36

Variation of the internal motion as the ramp angle is steepened for the PP10 to PP20 series of experiments. The observation point is 102.5cm from the end wall of the lock.

2.8   PROCESSES INVOLVED IN THE GENERATION OF A REVERSED DENSITY CURRENT

2.8.1 Reverse flow genesis at the ramp

The generation of a reverse flow density current can be summarized as follows:

(i) The head of the forward density current hits the ramp (Fig. 2.13)

(ii) A bulge of fluid forms around the foot of the ramp and is supplied by fluid both from the gradually collapsing head and also, from the tail of the flow as it reaches the ramp. Whilst this is occurring, motion in the reverse direction is initiated in the lower part of the body (Fig. 2.13).

(iii) Whilst fluid is still being supplied to the 'bulge' from the collapsing flow on the ramp, it begins to propagate back towards the lock, moving through the residual tail of the forward flow (Fig. 2.13).

The processes occurring at the ramp essentially, comprise of the gravitational collapse of a finite quantity of denser fluid into a two-layered fluid system. This mechanism is basically the same as that which generated the original density current. However, in this instance, the saline solution collapses into a 2-layered, miscible, fluid system in which the basal layer (saline solution) will have approximately the same density as the collapsing fluid ('bulge'). In addition, the generation of the reverse current involves the quasi-continuous supply of denser fluid from the ramp, rather than the instantaneous supply associated with the generation of the forward current.

2.8.2 Evolution of the reverse flow

2.8.2.i Recap of the evolution of the reverse flow

As the 'bulge' of fluid, generated at the ramp propagated, it continuously changed form. Initially, it was a diffuse, single elevated mass of fluid which subsequently broke down to two. These two masses still had pervasive mixing and often, distinct billows were observed. In addition, the reverse flow had a strong asymmetry (Type A, Plate 2.2a). This form passed through intermediate stages, with a progressive increase in symmetry and decrease in mixing, as it continued to propagate (Type B, Plate 2.2b). Finally, the reverse flow was apparent as a group of symmetrical, smooth waveforms (Type C, Plate 2.2c), which had no mixing at the upper interface. The profile between the waves could be both horizontal or slightly concave, but were always smooth with no mixing.

The reverse flows, apart from their distinctive forms, were associated with both an increase in the mean flow depth (d_1) and flow velocity (U_b) when compared with d_2 and U_t of the tail, into which it propagated. In fact, U_b of the reverse flow in the experiments described herein varied from 70 to 130% of U_h. Such characteristics, identified the generated reverse flow with an internal bore (a moving hydraulic jump).

As discussed in chapter 1.8, internal bores can be generated in the laboratory by a variety of mechanisms including: (i) the collapse of a denser fluid into a density layered fluid system (Wood and Simpson 1984); (ii) the intrusion of a gravity current along a boundary or density interface (Britter and Simpson 1981; Rottman and Simpson 1989), and (iii) by towing an obstacle along the base of a density layered fluid system (Rottman and Simpson 1989). Regardless of the method of generation, three general bore types can be generated, which type forms being determined by the ratio d_1/d_0. The ratio d_1/d_0 can be considered to be a measure of the bore strength.

Specifically, Rottman and Simpson (*op cit*) outlined the following rule-of-thumb for a bore propagating into a fluid system that was **stationary** upstream of the bore:

$1 < d_1/d_0 < 2$ -Bore has smooth, undular form
Type C (Fig. 1.23a)
$2 < d_1/d_0 < 4$ -Bore is undular, but some mixing due to shear instability occurs on the back of the waves Type B (Fig. 1.23b),

$d_1/d_0 > 4$ -Bore has the appearance of a gravity current
Type A (Fig. 1.23c).

As Figure 2.16 illustrated, the bore type formed was strongly dependent upon the ratio d_1/d_0 . The results of the experiments conducted by this author suggested the following modifications to the rule-of-thumb presented above:

$d_1/d_0 > 3$ type A bore
$2 < d_1/d_0 < 3$ type B bore
$1 < d_1/d_0 < 2$ type C bore

Thus, there was good agreement with Rottman and Simpson (*op cit*) for type C bore only. Several possibilities exist that could explain the discrepancies:

(i) The subjective nature of the definition of a type A bore,

(ii) Measurement of d_1 which for type A and type B (with a large component of type A) bores had a large measurement error associated with it (Appendix 2A), compared with the measurement of d_1 for type C which was always well defined

(iii) In the experiments conducted by this author, the bore propagated into a system that, in the basal layer, had an opposing velocity that opposed that of the bore. This could increase the shear across the upper surface of the bore, thereby inducing shear instabilities (ie. Type A bores) at lower values of d_1/d_0. ·

Although all of the points mentioned above may contribute to the variations, a combination of (ii) and (iii) would be most effective in generating the discrepancies observed as the determination of a type A bore is usually straightforward.

It may be expected that as the volume of fluid that accumulated at the ramp increased then d_1, and hence d_1/d_0, would become greater. As the width of the flume tank is constant, the height of the fluid that collected at the ramp can be taken as an approximate measure of the volume. When compared with the ratio d_1/d_0, the available data did not offer conclusive evidence for this supposition (Fig. 2.37).

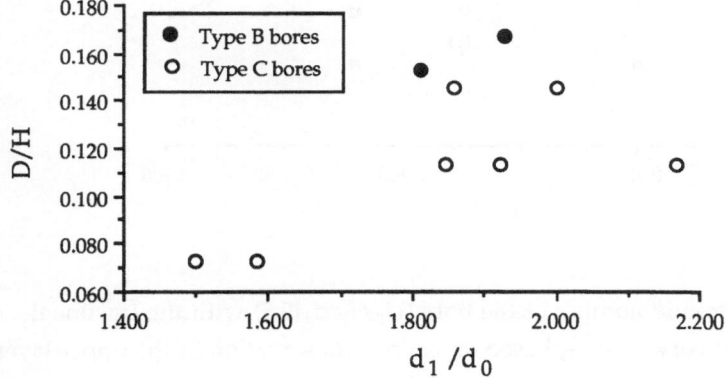

Fig. 2.37

Variation of flow strength with bulge height for the FE10 to FE20 series of experiments.

2.8.2.ii Generation of a bore by the gravitational collapse of a denser fluid into a layered system

Wood and Simpson (1984) generated bores by the quasi-continuous release of a finite volume of saline solution into a two-layered saline-freshwater system. The lower layer of this system had the same density as the volume to be released and both were a salt solution. The upper layer was freshwater and much deeper than the lower layer. Their experiments, which were designed to study the propagation of a bore into a two layered fluid initially at rest, agreed reasonably well with the theory of Chu and Baddour (1977) (see section 1.8, Fig. 1.23). The results of this authors' experiments show a similar scatter for bores of type B and C (Fig. 2.38). However, type A bores of had a measured velocity that was less than that, predicted by theory, for a specific fractional depth (d_1/H). This contrasts with the results of Wood and Simpson (1984) who found a good agreement between theory and measured velocities for type A bores.

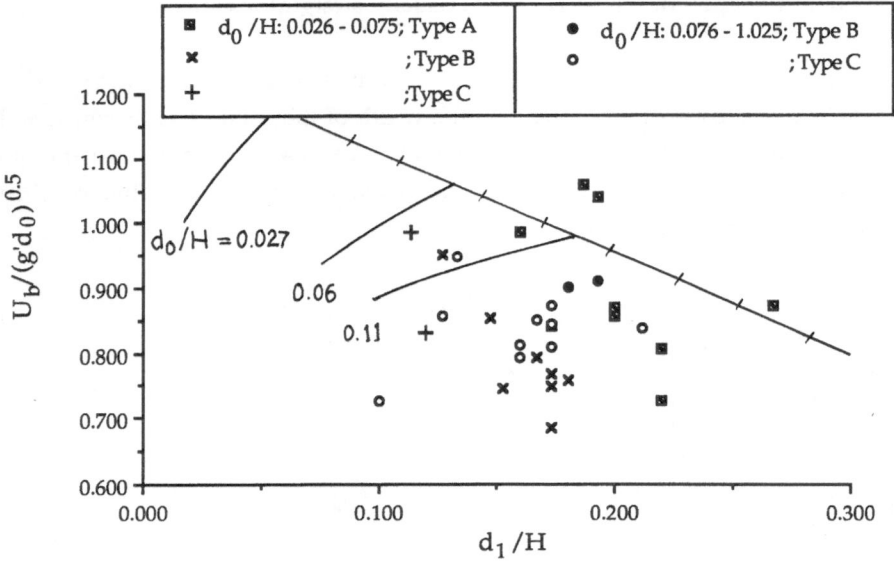

Fig. 2.38
The variation of the Froude number of the bore ($U_b/\{g'd_1\}0.5$) with the fractional flow depth (d_1/H). Theory: ———, based on energy conservation in the upper layer; ⊹⊹⊹, density current results (Simpson and Britter 1979). The experimental results from this study are grouped together dependent upon their d_0/H values.

The theory of Chu and Baddour (1977) was specifically for the case of a bore advancing into two layers that were initially at rest. This was complied to in the work of Wood and Simpson (*op cit*), however, for the experiments described herein, the bore advanced into non-stationary layers. As described in 2.7.8, the velocity within the tail can be as much as 34% of the crestal velocity of the wave. It is therefore, not unreasonable to propose that the velocity of the tail, which is often declared as being negligable and thus dismissed, is a control on the velocity and perhaps on the form of the reverse flow.

A further discrepancy between this authors' results and theory may have resulted from the method of generation of the bores. Theory and previous experimental, consideration of hydraulic jumps involved the use of the least complicated model in which the variables had been minimised and could be well controlled. The generation of bores via the reversal of density currents from various low-angled ramps, as applied by this author, is a far more complex situation and it is therefore, not suprising to find that discrepancies occur.

2.8.2.iii The generation of bores by the intrusion of a density current underneath a layered fluid system.

If the basal layer was slightly less dense than the fluid accumulating at the ramp, then the reverse flow will intrude along the bottom boundary as a density current, generating a bore on the interface above (chapter 1.8). According to Rottman and Simpson (1989), when $d_1/d_0 < 4$ the bore moves ahead of the density current, periodically removing portions of the density current head. Such a bore, in its early stages of development would have a distinct, front wave with a wavelength and amplitude that was considerably larger than that of the succeeding waves (Rottman and Simpson 1989). The occasional occurrence of a nose and lobes and clefts at the front of the reverse flow in the experiments described here, may suggest the occurrence of a density current underneath the basal layer, that became affected by friction at the base of the flow. This was certainly not a ubiquitous feature and where it occurred, it was in the front wave rather than those behind, as theory requires (Wood and Simpson 1984).

The formation of a bore by the intrusion of a density current along a boundary beneath a layered system is adequately described by the fluid regimes generated by towing a streamlined obstacle at the bottom of a two-layer fluid (Rottman and Simpson 1989). To utilise the theory, the density current height and speed are substituted for the equivalent values of the obstacle. In the authors' experiments, no head could be distinguished and therefore, the relevant values of the bore had to be used. The results for the bore fitted reasonably well with theory, and occurred within the field of partially blocked flow (Fig. 1.20).

Thus, the reversal of a density current from a low-angled ramp in a rectangular flume tank generates a bore, the form of which is dependent upon d_1/d_0. In the experiments described in this chapter it was not possible to show, conclusively, that the reverse flow was either more or less dense than the tail of the flow. As any density difference that could occur would be very small it may be that aspects of both hypotheses occur and the resultant bore is a 'combination' flow.

2.8.2.iv Comparison of the generated bores with density current and internal solitary wave theories

When $d_1/d_0 > 4$ a bore has the appearance of a density current although, when $d_1/d_0 > 2$ the bore has velocities that are described by the velocity of advance of a density current (Wood and Simpson 1984). Comparison of measured velocities with velocities calculated both for a density current head (equation 1.4) and a bore propagating into stationary layered fluid system (equation 2.3) verify this statement (Fig. 2.39).

$$U_b /(g'd_0) = \{0.5(d_1/d_0)(1 + d_1/d_0)\}^{0.5} \qquad\qquad (2.3)$$

(Rottman and Simpson 1989)

The bore theory overpredicts the propagation velocities by approximately 50% whereas, density current theory describes the observed velocities almost perfectly.

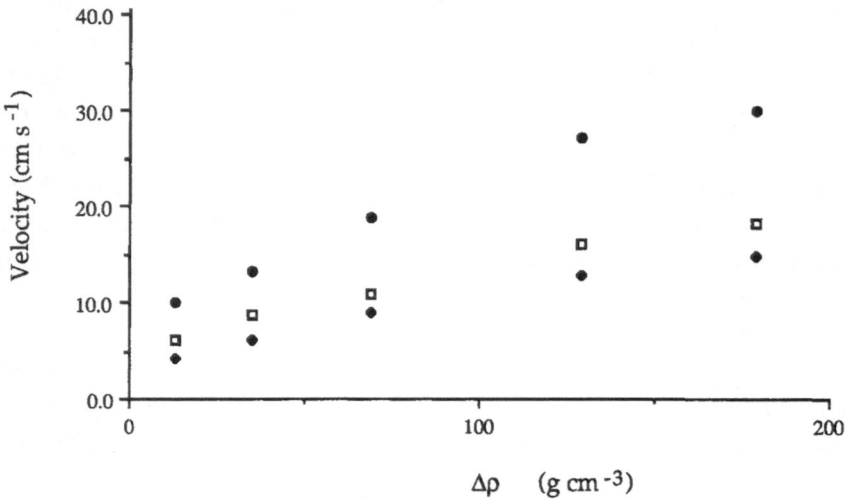

Fig. 2.39

Comparison of the measured velocities of reverse flow with density current and bore theory for the S15 series of experiments. □ represents the measured values, ◆ those calculated using the Keulegan (1957) relation for the propagation of a density current (equation 1.4) and ● the velocities calculated for the propagation of a bore into a system initially at rest (equation 2.3).

The wave trains associated with a weak undular bore are thought to be cnoidal in origin (Benjamin and Lighthill 1954) of which solitary waves are one specific form (chapter 1.9). Theory and observations have indicated that in time, the waves associated with an undular bore will tend towards solitary waves (Benjamin and Lighthill *op cit*). As individual waves within the wave train become more solitary in nature, the wave train gradually becomes stretched out in the troughs relative to the peaks, ie. it develops an asymmetry.

The groups of waves associated with undular bores in the experiments discussed herein, do become more separated with time, this effect being most obvious at lower values of $\Delta\rho$ (Plate 2.4). As such, the waves tend towards solitary waves and should be described by

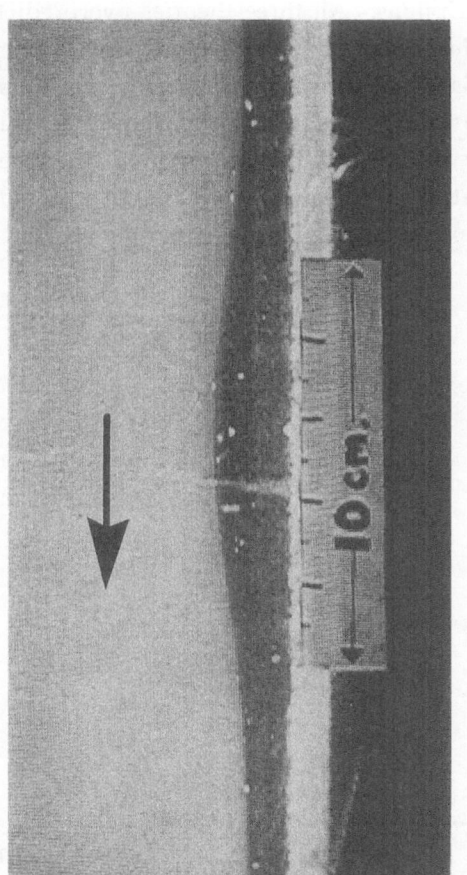

Plate 2.4

Vertical view of the profile of an internal solitary wave that occurs at the front of a type C bore. The flanks of the waves approach the horizontal either side of the crest.

The arrow indicates the direction of the movement of the waveform.

either the theory for shallow water (KdV), deep water (BDO) or finite depth internal solitary waves (equations 1.27, 1.34 and 1.41 respectively, Fig. 2.40).

According to the categorisation of Koop and Butler (1980) (see 1.9.3) the finite depth theory should be the best description as L/H >>1 and h/H <<1. However, the finite depth equation provides the least agreement with the measured results with both the BDO and KDV theories providing better descriptions. All three theories overpredict the measured values. This is not unexpected as the theories envisage waves propagating in an inviscid fluid system in which there is no miscibility between the layers. The diffuse nature of the boundaries within a miscible, saline solution-fresh water system will generate greater flow thicknesses for a particular velocity resulting in theory overpredicting the velocities. In addition, viscosity and friction (that are not accounted for in an inviscid analysis) will serve to generate velocities below those predicted by theory for the specific measured flow dimensions.

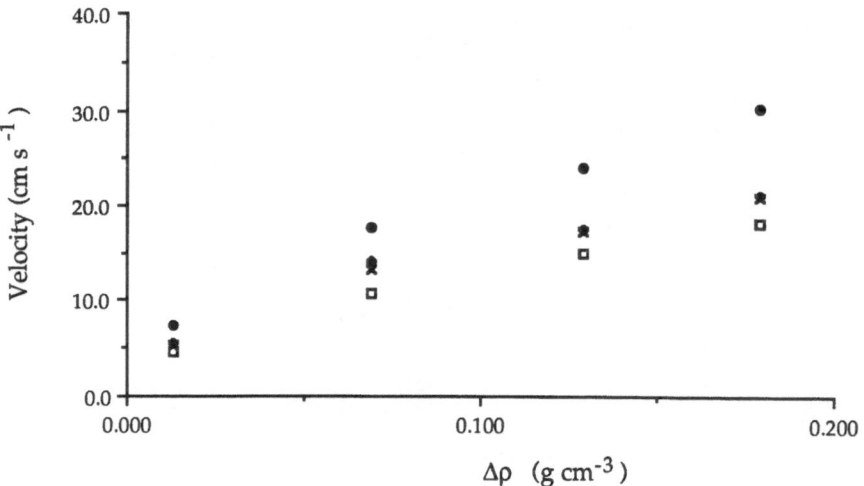

Fig. 2.40
Comparison of the measured velocities (□) of the reverse flow with the theories for deep(x), shallow (♦) and finite depth (●) internal solitary wave propagation for the FE15 series of experiments.

Comparison of the velocities of the BDO and KDV theories does not allow distinction between these two theories to be made in these laboratory experiments. However, in the deep ocean the crestal velocities of the waves on the upper surface of an undular bore should, in time, approximate to the BDO deep water wave equation.

The most conclusive evidence for the waves generated on the upper surface of a weak undular bore being solitary in nature, came from experimental observations rather than theory:

(i) As described in earlier sections, the waves became increasingly spread out with both time and distance until they appeared as independent, solitary wave features.

(ii) After reflection from the vertical end wall of the flume tank and then, subsequent interaction with the other waveforms in the group the waves emerged unchanged. This indicated that the waves, particularly in the FE10 to FE20 series of experiments, were **solitons** by the time they had travelled from the ramp to the lock.

In conclusion, the waves associated with the undular bore stage of the flow do approximate to internal solitary waves. However, there is a continuum from a more general cnoidal wave train to a group of solitary waves, with the solitary wave form being associated with the weakest bore.

2.9   THE INCIDENCE OF UNRESTRICTED DENSITY CURRENTS UPON PLANAR RAMPS

The majority of the present study concentrated upon investigations of flow in a rectangular tank, thereby considering the effects of reversal of an essentially 2-dimensional current. Geologically, it is important to determine the nature of the current generated by the incidence of a 3-dimensional, radially spreading current upon a ramp set at an angle to the axis of symmetry. In order to examine this situation, a series of experiments were conducted in the square tank (Fig. 2.2). Upon releasing the current from the exchange lock, a restricted flow developed in the channel which then spread radially upon reaching the channel mouth (Plate 2.5).

The incidence of a radially spreading current upon a 20° ramp positioned at right angles to the axis of symmetry of the spreading flow generated three distinct waveforms which moved back towards the lock (Fig. 2.41). The crestal axes of the waveforms were straight and ran parallel to the strike of the ramp.

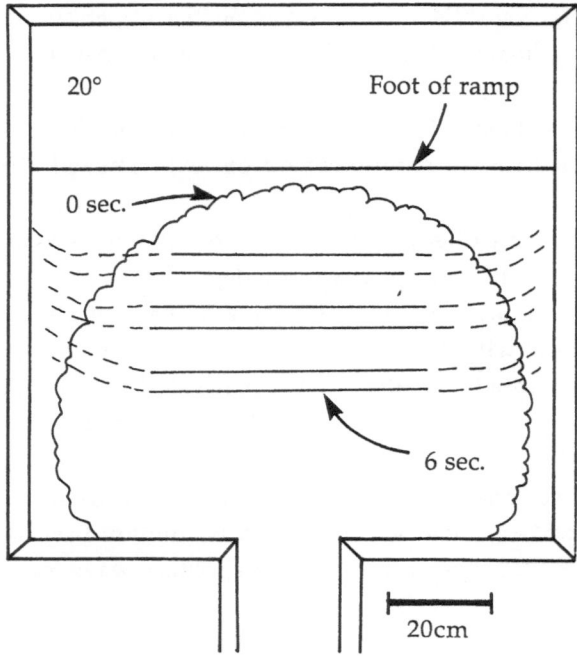

Fig. 2.41

Plan view of the generation of a weak undular bore by the incidence of a radially spreading density current upon a ramp. The strike of the ramp was at 90° (in a horizontal plane) to the axis of symmetry of the original current. Three smooth waves were formed and are illustrated approximately 6sec (t = 6) after the original density current first impinged on the ramp (t = 0). The curvature of the crestal axes near the edge of the flume tank is a result of side-wall friction.

Plates 2.5, 2.6 and 2.7

Plan views of the incidence of an unrestricted, radially spreading density current upon a 20° ramp, whose strike is 45° to the axis of symmetry of the forward flow. Arrows indicate the direction of movement of the bore. Times (t) are in seconds after Plate 2.5a.

Plate 2.5

(a) t = 0s. Spread of a radial forward-moving density current.

(b) t = **1**s. The forward-moving density current is just incident upon the ramp. A darker mass of fluid has begun to accumulate at the foot of the ramp.

Plate 2.6

Waveforms with crestal axes that are approximately parallel to the strike of the ramp move away from it. The interaction of waves generated at the ramp with those generated at the vertical wall of the flume tank (right-hand side), results in curved crestal axes.

(a) t = 2s.

(b) t = 3s.

Plate 2.7

Two distinct waveforms and a third that is less well defined have formed.

(a) t = 4s.

(b) t = 5s.

Plate 2.5

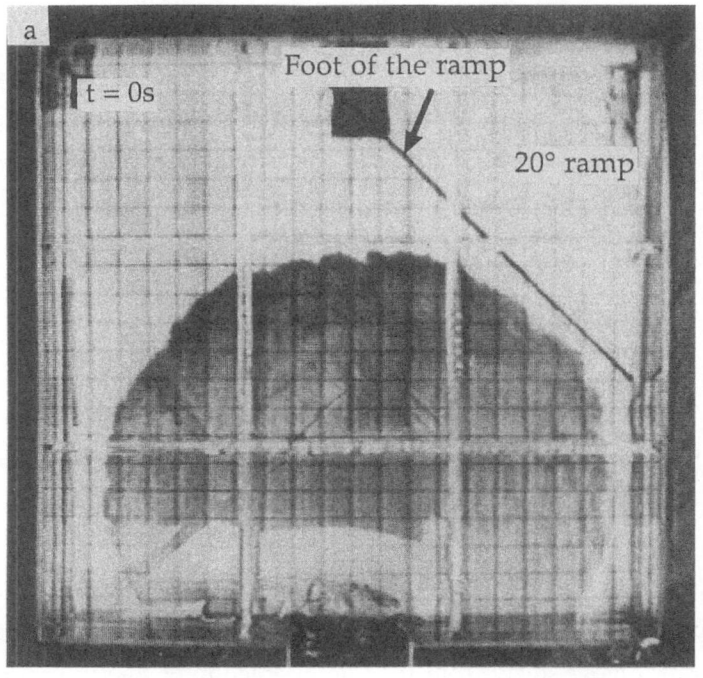

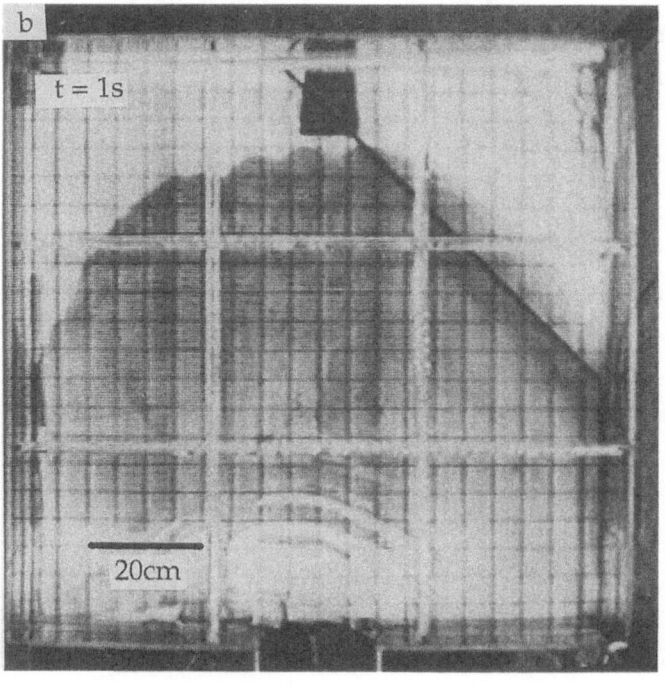

Plate 2.6

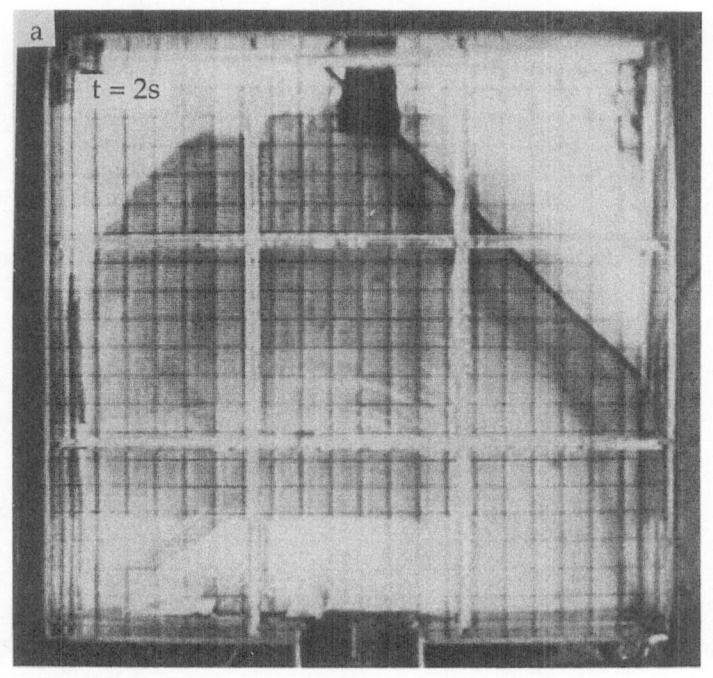

a

t = 2s

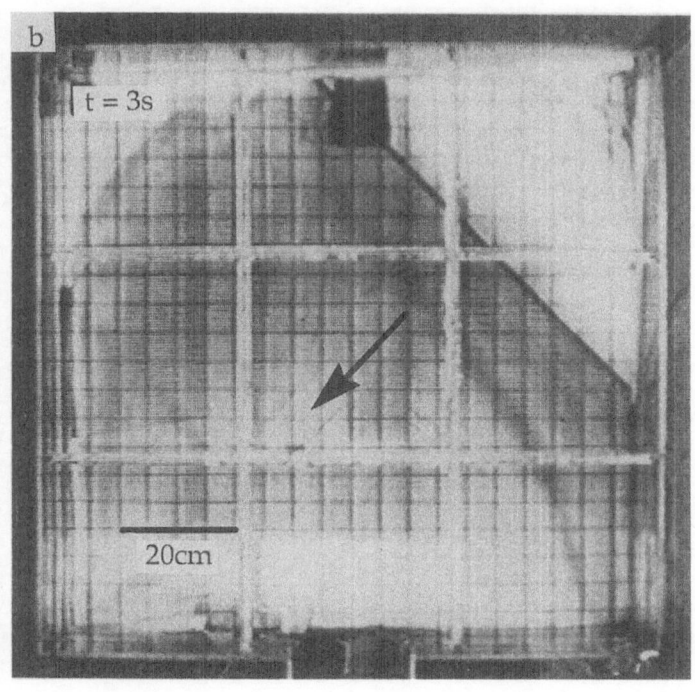

b

t = 3s

20cm

Plate 2.7

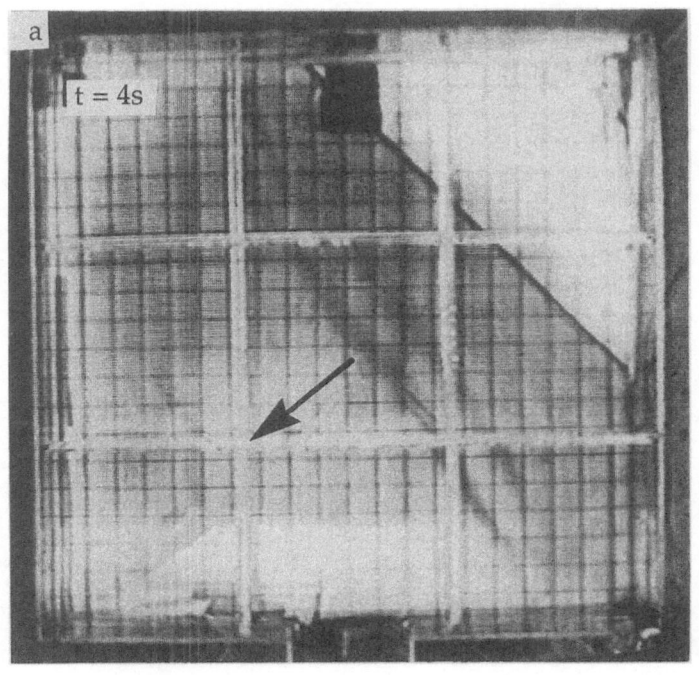

t = 4s

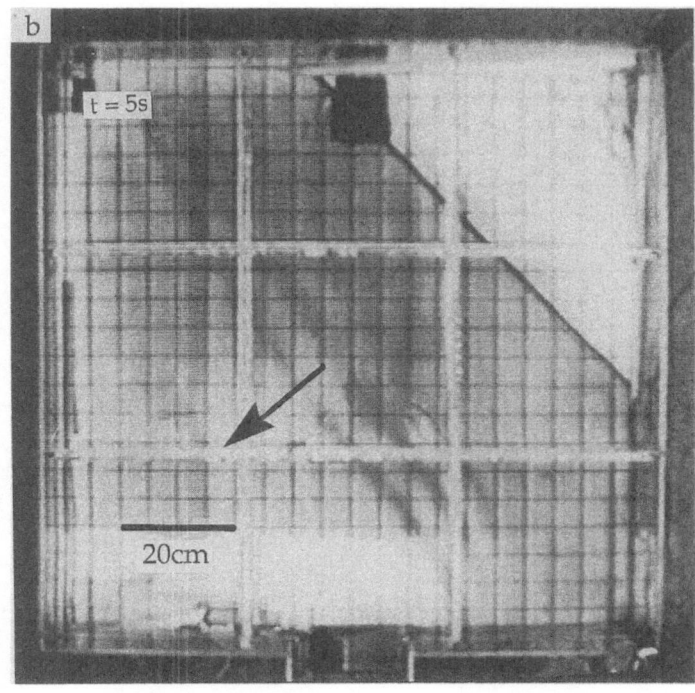

t = 5s

20cm

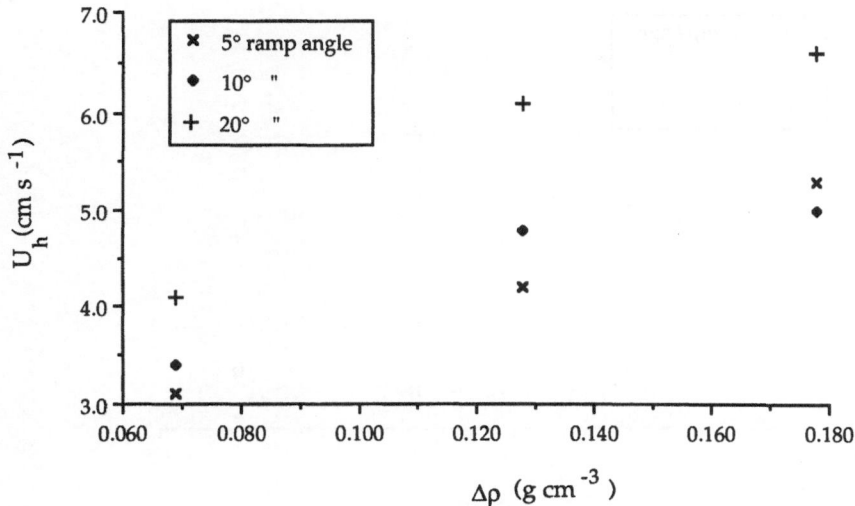

Fig. 2.42

Variation of the velocity of the first waveform (U_h), with density difference ($\Delta\rho$) for the T5 to T20 series of experiments. The velocity of advance of the head of the current is taken as being the velocity of the rate of advance of the axis of symmetry of the current.

In a more detailed study, reflection was investigated from ramps placed at 45° to the axis of symmetry of the spreading current. Once again, the incidence of a forward moving current upon the ramp resulted in the generation of straight-crested waves, with crestal axes parallel to the strike of the ramp, that moved away from the ramp (Plate 2.6, 2.7). Neither lobes, clefts or billows occurred.

As in the rectangular tank, U_b increased as U_h, $\Delta\rho$ and the angle or the ramp become greater (Fig. 2.42) although, when $\Delta\rho = 0.179$g cm^{-3}, U_b is 5.3cm s^{-1} at 5° compared with 5.0cm s^{-1} at 10°. This discrepancy can be accounted for by the measurement errors associated with the measurement of the velocity.

More specifically, as the ramp became more steeply inclined, U_b approached U_h. For example, from the 5° ramp U_b is 42 to 44% of U_h compared with 55 to 62% from a 20° ramp (Fig. 2.43). The relationship between U_h and U_b is apparently independent of the density of the saline solution.

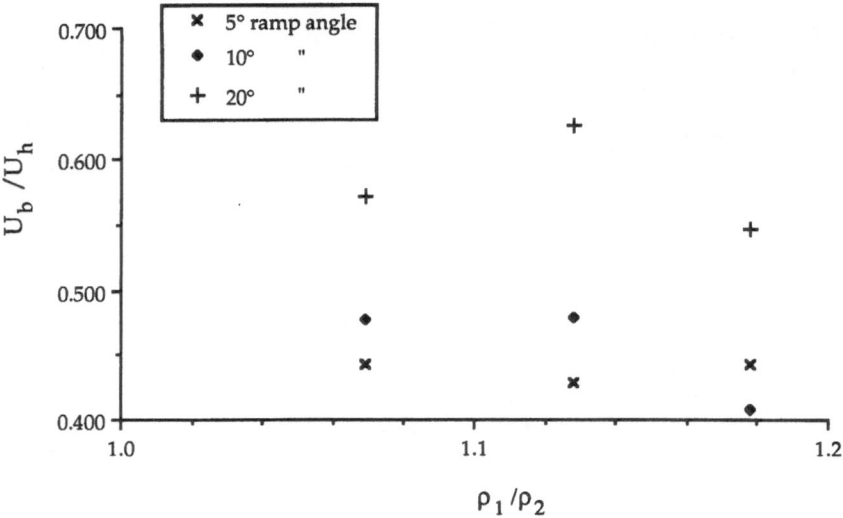

Fig. 2.43
Variation of U_b/U_h with the density ratio ρ_1/ρ_2 for the T5, T10 and T20 series of experiments.

The forward current reached the bottom right hand corner of the ramp approximately one second before being incident upon the upper right hand corner (Plate 2.4). Thus, it may be expected that a generated wave would form sooner at the end of the ramp nearest the source and as a result would have a curved crestal axis. However, the crests of the waves generated at the ramp ran approximately parallel to the strike of the ramp (Plate 2.6, 2.7). Apart from the timing of incidence of the forward current on the ramp, straight crested waves are unexpected as

(i) U_h varies from a maximum parallel to the symmetry axis of the current to a minimum at 90° to this,

(ii) U_b of the wave forms should vary from a maximum where the generating forward current travelled up the dip slope, to a minimum where it propagated up the smallest apparent slope.

The generation of reverse currents at the vertical sides of the walls also resulted in symmetrical waveforms. These sidewall waves interacted and merged with those from the ramp producing slightly curved crestal axes (Plate 2.7). Complex interactions of waves from the edges of the flume tank rapidly complicated and masked the subsequent flow from the ramp preventing a more detailed analysis of wave evolution.

As described in chapter 1 (see section 1.9.7), Maxworthy (1980) generated similar features by the three-dimensional collapse of a denser fluid into a layered medium. He described the generated waveforms as internal solitary waves stating that they were an ubiquitous feature of his experiments.

2.10   THE INCIDENCE OF A RESTRICTED SEDIMENT SUSPENSION CURRENT UPON A   PLANAR RAMP

The main body of the experimental study has used salt solutions as the denser fluid. When considering general flow characteristics this method has many advantages over the use of sediment suspensions. For example, there are no complicating effects, such as pseudo-plastic behaviour, to consider when increasing the concentration of the saline solution; settling velocities do not need to be considered and it is more easy to visualise the internal flow properties. However, it is obviously important to determine whether or not the features generated by the incidence of salt solution density currents upon a ramp, will also occur with sediment suspensions.

The experimental modelling of the flow generated when a suspension density current was incident upon a ramp, required the use of a current that would not peter out too soon. However, maintenance of the flow under laboratory conditions was the major problem encountered when either artificial or natural, **real-scale** sediment provided the density excess in a suspension density current (section 1.5), although, the use of very fine-grained sediment has had more success. It should be pointed out, however, that at very fine-grained sizes, providing substantial flocculation does not occur, the fine grained sediment, essentially, mimicked a salt solution density current (Riddell 1969). This is because the settling velocity was long with respect to the duration of the current.

An experiment early in the program indicated that a reverse flow consisting of waveforms, would indeed form whilst using a sediment suspension in water. The sediment in this instance was surgical Plaster of Paris, the setting properties of which precluded any further study with this medium. As it was very difficult to ensure that the flume tank was entirely free from salt, flocculation of clays was a problem. Finally, when silica flour suspension in fresh water was used, a flow was generated that could be analysed without difficulty.

An experiment was set-up to replicate experiment FE20/1.068. Prior to hitting the ramp, U_h of the forward current (Plate 2.8) was approximately 5.8cm.s^{-1} (66% of U_h at the equivalent stage in FE20/1.068). The maximum value of h_1 was the same as in FE20/1.068 (ie. 6.1cm). Behind the head of the suspension current a layer of sediment approximately 2cm thick was rapidly deposited as the current propagated, reducing the effective height of the head. The head of the current travelled 11.1cm (vertically) up the ramp, ie. 74% of the total flow depth, before petering out.

Plate 2.8

(overleaf)

Vertical view of the flow generated by the incidence of a restricted silica flour suspension current upon a 20° ramp. Time (t) is in seconds after Plate 2.8a. The arrows indicate the direction of motion of the bore.

(a) t = 0s. The head of a forward-moving turbidity current.

(b) t = 82s. The first waveform at the front of the bore. Note the thicker flow thickness that follows the waveform compared with that prior to the arrival of the bore.

(c) t = 85s. The first and second waves of the bore.

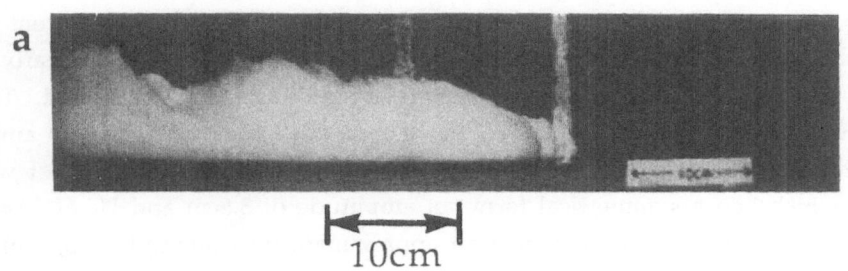

10cm

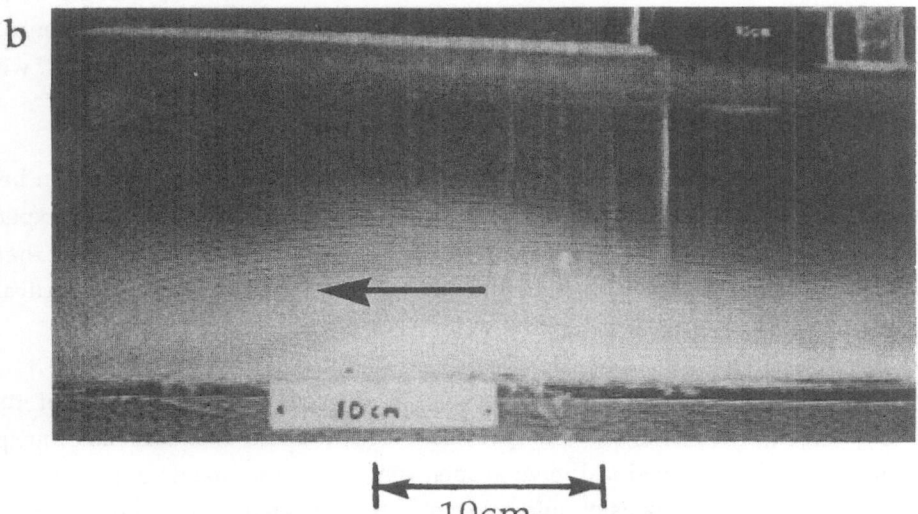

10cm

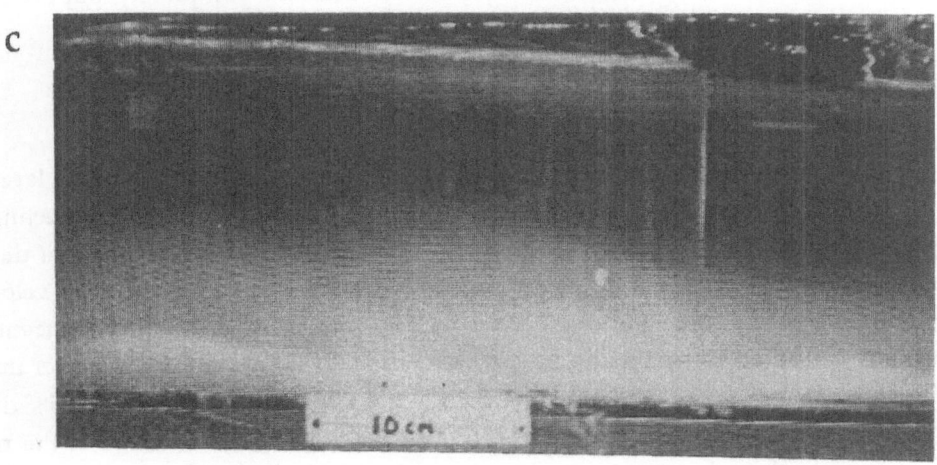

Plate 2.8

The opaque nature of the suspension masked the features of the reversed current until it was 1.00 to 1.25m from the foot of the ramp. At this point three waves were clearly visible (Plate 2.8). The first and third waves were larger in amplitude than the second. The first waveform had an asymmetrical form with the steeper side facing upstream, an amplitude of approximately 3.7cm and U_b of about 5.6cm s^{-1}, in contrast with the first wave in FE20/1.068 which had a symmetrical form, an amplitude of 3.8cm and U_b of 10.4cm s^{-1}. The reverse flow suspension current had an approximate flow strength of 3, which was borne out by the asymmetrical form of the waves. Qualitatively, the second and third waves were both smaller than the first, however, the third wave had a larger amplitude than the second (Plate 2.8). The first wave was distinct from the succeeding ones, there being approximately 30cm of horizontal flow before the second and third waves, which in contrast were closely spaced.

Reflection from the vertical end-wall of the flume tank resulted in an interaction between the waves from which they emerged apparently unchanged. The silica flour experiment petered out before travelling half the distance back along the flume tank towards the ramp. This is in contrast to FE20/1.068 in which the reverse current from the vertical wall reflection reached the ramp again, the first wave having a velocity of 5.0 cm s^{-1}.

In essence, the sediment suspension experiments agreed with the results of the salt solution density currents although, as previously mentioned, this was not unexpected when using a very fine-grained sediment suspension. The experiments demonstrated that, fallout of sediment from suspension and other effects associated with sediment suspension flows, could alter the ranges of d_1/d_0 over which the different bore types occurred. However, the effects of the sediment suspension did not completely suppress the formation of bores and their associated waveforms.

2.11 THE GENERATION OF BORES DURING FLOW STRIPPING

As previously discussed, the ramp angle had some influence in determining how large the bulge of fluid that accumulated at the ramp would be (therefore potentially affecting the strength of the bore and its velocity). Providing that the forward density current did not flow over the top end of the ramp, the generated reverse flow should have lower velocities and smaller bore strengths, as the angle of the incline upon which the forward current was incident, was decreased. Stripping off part of the current, by allowing it to flow over the top end of the ramp, as in the S5/1.068 to S5/1.178, FE5/1.068 to FE5/1.178 and the CD5 series' of experiments, should result in a further decrease in the volume of the bulge at the ramp. As a consequence, the subsequent decrease of velocity and bore strength should be greater than would be expected for just a decrease in the ramp angle.

An exploratory study showed that when the head of the current flowed off the end of the ramp the generated reverse flow usually consisted of a group of smooth, symmetrical waves. However, when H = 5cm, $\Delta\rho$ = 0.069 g.cm^{-3} and the foot of the ramp was 370.5cm from the endwall of the lock, no waveforms were visible and the reverse flow consisted of a barely discernable motion back towards the lock. The waves generated in the S5/1.068 to S5/1.178 series of experiments were smooth and symmetrical. The velocity of the first wave increased with increasing $\Delta\rho$ and was not anomolously less than for the flows at higher ramp angles where flow stripping did not occur (Fig. 2.44).

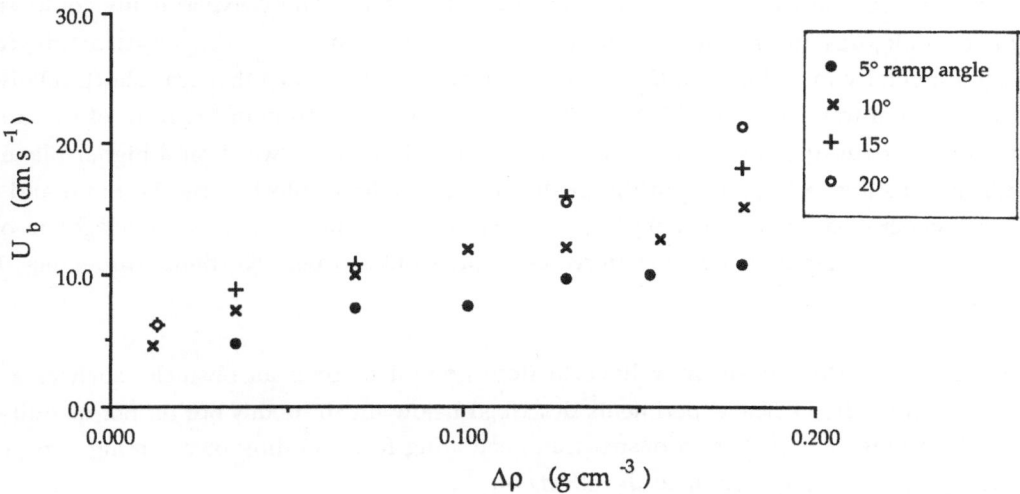

Fig. 2.44

Variation of the bore velocity (U_b) with the density difference ($\Delta\rho$) for the S5 to S20 series of experiments.

There is evidence for the difference in velocities being greater at higher values of $\Delta\rho$. This was to be expected, as at higher values of $\Delta\rho$ a larger proportion of the forward current flows over the end of the ramp and hence, the volume of fluid that accumulated at the base of the ramp should be even less. The bore strength and amplitude were also appreciably lower than in corresponding flows at higher ramp angles where flow stripping did not occur. Again the difference was not anomolously low. In addition, the increase of bore strength with increasing ρ_1/ρ_2 was not apparent when flow stripping occurred. As more of the flow went over the end of the ramp, the corresponding reduced bulge size must have been similar at all values of $\Delta\rho$, hence maintaining approximately constant bore strengths.

In the FE5 series of experiments the reverse flow was significantly different. The reverse flow was characterised by one large waveform that in FE5/1.128 and FE5/1.178 were preceded by 3 and 2 much smaller waves respectively. Where the velocities of the large waves could be measured they were slightly larger than the corresponding wave velocities in the FE1.068 series of experiments (Appendix 2B).

Flow stripping, therefore, does not inhibit the generation of bores from the incidence of a forward moving density current on a short ramp. It may, however, modify the flow to some degree, the nature of which is as yet unexplored. These experiments could easily have analogues in the marine environment. For example, these experiments could experimentally model a turbidity current that moved up a scarp that was short, relative to the potential travel distance of the current. A significant portion of the front of the current would over-run the end of the scarp and continue flowing forward on a higher plane. In addition to this, a large proportion of the current could be blocked by the ramp and as a result generate a reverse or deflected flow. Such a mechanism has been envisaged in order to explain the depositional characteristics of the Contessa bed, Northern Appenines, Italy (Ellis 1982).

Alternatively, the current may have to flow around or over an obstacle, such as a sea-mount, in the flow. The experiments described below illustrate this but do not quantify the modifications, if any, that an obstruction generating flow splitting or stripping may incur. An exploratory study examined the effects of:
   (i) a planar obstacle that extended the full width of the rectangular flume tank, and
   (ii) a hemispherical obstacle in the square tank.

In the rectangular tank, a saline current of initial $\Delta\rho = 0.029g$ cm^{-3} was released into water of total depth 15 cm. The obstacle was placed 289cm from the end wall of the lock and was approximately $0.25h_1$, $0.5h_1$ and h_1 high. In all instances, when the forward current hit the obstacle a bulge of fluid formed in the stagnant area in front it. This bulge then moved back and was affected by the processes described in 2.7.2. The evolution of the flow when incident upon an obstacle of height approximately equal to h_1 is illustrated in Figure 2.45a to f.

The bore strength progressively increased from < 2, when the obstacle height was less than h_1, to > 2 when it was approximately equal to h_1. This increase in bore strength was indicated by the occurrence of mixing on the back of the generated waveforms when the obstacle height was equal to h_1. The increase is attributable to the increase in the volume of the bulge at higher obstacle heights.

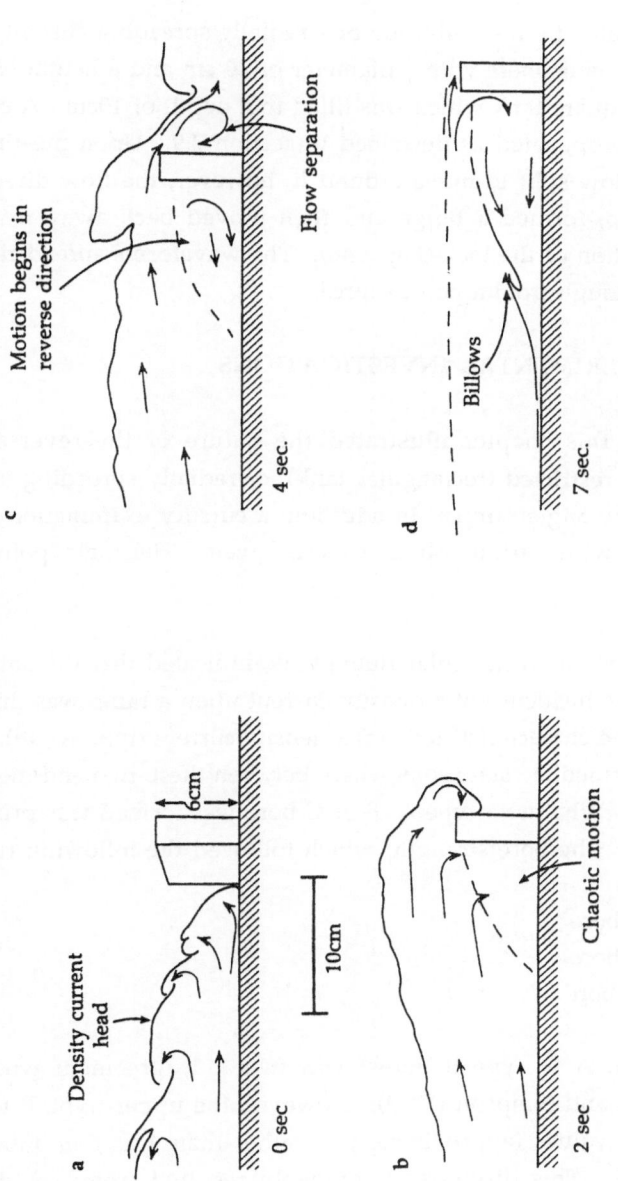

Fig. 2.45

Generation of an undular bore from the incidence of a density current upon a vertical obstacle that is oriented at 90° to it.

An undular bore was also generated by the deflection of a radially spreading current in the square tank. In this instance, a hemisphere with a diameter of 20 cm and a height of 4 cm was placed in the centre of the square tank which was filled to a depth of 10cm. A current of initial $\Delta\rho$ was released and propagated as described in section 2.9. Upon meeting the obstacle, the outer part of the flow split to move around it, however, the flow directly in front of the obstacle banked up, formed a bulge and then moved back away from the hemisphere in the general direction of the lock (Fig. 2.46). The waveforms spread radially away from the hemisphere as though from a point source.

3.12 SUMMARY OF THE EXPERIMENTAL INVESTIGATIONS

The experiments described in this chapter illustrated the nature of the reverse flow generated by the incidence of a restricted (rectangular tank), or radially spreading (square tank) density current upon a low-angle ramps. In addition, a cursory examination of the interaction of density currents with various obstacles was given. The main points are summarised below:

(i) The experiments conducted in the rectangular flume tank, indicated that the nature of the reverse flow generated by the incidence of a density current upon a ramp was that of a bore. This generated bore had the characteristics of: (a) a density current (type A), (b) group of waves (type C) or, (c) an intermediate state somewhere between these two end-members (type B), superimposed upon it. Whether a type A, B or C bore was formed was primarily dependent upon the ratio d_1/d_0 (the bore strength) which followed the following rule-of-thumb:

$$1 < d_1/d_0 < 2 \qquad \text{type C bore}$$
$$2 < d_1/d_0 < 3 \qquad \text{type B bore}$$
$$d_1/d_0 > 3 \qquad \text{type A bore}$$

(ii) The lower boundary (type A to type B bores) was in good agreement with that previously outlined by Rottman and Simpson (1989). However, the upper (type B to type C) boundary occurred at lower values (approximately 3, rather than 4 as determined by Rottman and Simpson *op cit*). This discrepancy probably resulted from: (a) slightly different definitions of the form of a type C bore, and (b) the effect of the opposing velocity in the tail being significant, inducing shear instabilities and a density current form at lower values of d_1/d_0.

(iii) If the 'bulge' of fluid that collects at the ramp was the same density as the tail of the forward density current, then the bore was generated by the surge of the bulge back towards

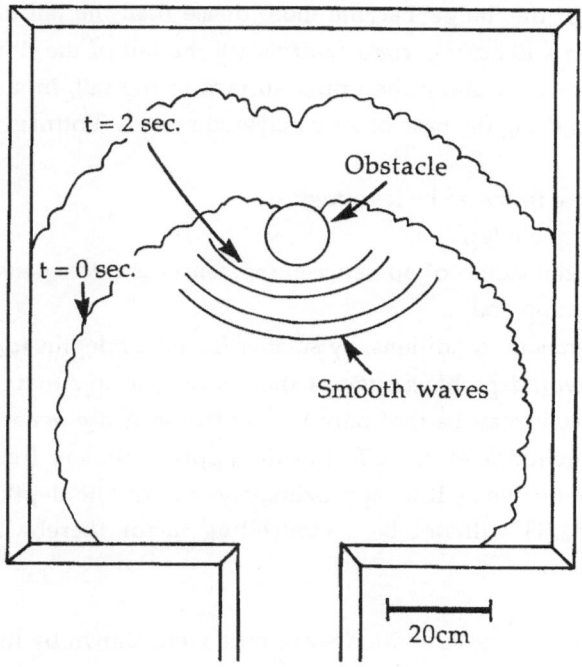

Fig. 2.46

Plan view of the generation of a weak undular bore by the flow of a radially spreading density current around a hemispherical obstacle. Two smooth, concentric waves were generated and are illustrated 2sec (t = 2) after the forward current first enclosed the obstacle (t = 0). The waves propagated back towards the channel.

the lock. Conversley, if the 'bulge' became more dense than the tail of the flow, then the bulge would intrude as a density current underneath the tail of the flow. In this instance, the bore would be generated along the upper surface of the tail, in a manner akin to the dragging of an obstacle along the base of a two-layered system (Rottman and Simpson 1989).

(iv) Stronger bores were found to be favoured:
     (a) by higher ramp angles,
     (b) by greater initial values of $\Delta\rho$ between the denser and less dense fluids,
     (c) closer to the ramp and,
     (d) under experimental conditions, by smaller fractional depths (d_0/H).
The latter condition, would perhaps suggest that as in marine situations, the ratio d_0/H will tend towards zero, it may be that only bores of type A are generated. On the other hand, in the marine environment, the effect of the upper boundary (reflected in the use of H as a variable), is negligable as it is, approximately, an 'infinite depth' situation. In such instances, the ratio d_0/H will not be a controlling factor thereby not precluding the formation of type C bores.

(v) The group of waves associated with a weak bore were shown by theory (Benjamin and Lighthill 1954) to tend towards solitary waves. The form of the waves generated in these experiments agreed with this although, the agreement between solitary wave theory and the waves of these experiments was not very good. This poor agreement of calculated velocities with theory was not unexpected, as workers using controlled experimental conditions (eg. Koop and Butler 1981) also encountered similar problems.

These experiments have shown empirically, that internal solitary waves are produced when a density current is incident upon a ramp or obstacle. In the marine environment, the internal solitary waves associated with the reversal or deflection of turbidity currents will have the form and properties of the BDO deep water waves.

(vi) The incidence of radially spreading currents on ramps that were positioned at an angle to the direction of propagation of the current, also generated bores. These experiments only generated type C bores, whose crestal axes ran parallel to the strike of the ramp, although the values of $\Delta\rho$ used were adequate to generate type A and B bores in the rectangular flume. As the radial spread of the current involved a more rapid decrease in the velocity of advance of the head (more rapid dilution?) and hence, at the ramp, a lower effective $\Delta\rho$ value was applicable than if the current had remained in a rectangular channel.

(vii) It was shown that the interaction of either a restricted or radial spreading current with obstacle generated bores. When the obstacle height was approximately equal to the head height of the forward current it was clear that the bore was generated by the intrusion of a density current **underneath** the residual tail of the forward flow. In all of the other experiments (with ramps as well as obstacles), the generation mechanism was more akin to the release of denser fluid into a layered system in which the lower layer had the same density as the fluid being released. However it is likely that components of both mechanisms occurred.

(viii)The experimental density currents were similar to the characteristics of the flow of fine-grained turbidity currents. It is very probable that the flow properties generated in the experiments described in this chapter could occur, via similar mechanisms, within marine turbidity currents. However, as the exploratory experiments with sediment suspension currents showed, some modifications of the reverse flow as a result, primarily, of sediment fall-out will occur.

CHAPTER THREE

A MODEL FOR THE INCIDENCE OF GRAVITY CURRENTS ON OBSTRUCTIONS
TO FLOW AND A DISCUSSION OF ITS APPLICABILITY TO MODERN AND
ANCIENT TURBIDITE SYSTEMS

We shall not cease from exploration
And the end of all our exploring
Will be to arrive where we started
And know the place for the first time

T. S. Eliot (1888-1965)

3.1 INTRODUCTION

This chapter presents a model which describes the consequences of the incidence of density currents upon low-angle ramps and obstacles. The model assumes that the density of the generated flow is equal to or greater than that of the residual flow of the forward current. The model is expanded to the incidence of a suspension current (turbidity current) upon obstructions of the forward propagation of the flow. Subsequent to this, data from natural turbidity currents and experimental relationships are used to estimate potential magnitudes of the generated flow. The plausibility of the model as a flow mechanism capable of generating diagnostic sedimentary structures that can be identified in both recent and ancient turbidites is considered. This point is illustrated by consideration of turbidites within the Windermere Group of the Lake District, northern England.

3.2 A MODEL FOR THE INCIDENCE OF DENSITY CURRENTS
UPON FLOW OBSTRUCTIONS

3.2.1 Introduction

Essentially, the incidence of a density current, whether restricted or not, upon a ramp or obstacle will generate a bore. This bore is not necessarily a reflection, deflection or even a reversal *per se* . It must be considered as a flow phenomena which moves in a direction that opposes the forward current to some degree up to a maximum of 180°. The form of the bore is governed by the geometry of the obstruction to the flow that produced it. The bore may consist of all or only part of the original flow with the amount of fluid that is involved being dependent upon both the properties of the original flow and the nature of the obstruction.

3.2.2 A simplistic model

Let us consider the most simple case, that of a restricted current hitting a ramp whose strike is orthogonal to the propagation direction of the density current (Fig. 3.1). As the density current impinges upon the ramp, the head, mixing region and body begin to move up it. Whilst this occurs the current undergoes a continuous deceleration as the effects of friction, and more importantly, gravity becomes prevalent.

An elevated mass of fluid then collects at the foot of the ramp (Fig. 2.13). This mass is fed by dense fluid from the collapsing head, body and the tail higher up on the ramp. In addition, fluid is also supplied from the tail of the current that is still moving in the forward direction towards the ramp.

Reverse motion within the lower part of the body occurs almost immediately the forward flow begins to propagate up the ramp. Reverse flow is not obvious, however, until the elevated mass of fluid that accumulated at the foot of the ramp begins to propagate in the reverse flow (Fig. 2.13). As the bulge (surge) moves back its behaviour is dependent upon whether the density of the surge is greater than or equal to the density of the tail of the forward flow.

When the density of the surge is greater than the density of the tail, it propagates as a density current **underneath** the residual forward flow. The form of the density current is very similar to that of the original forward flow having a distinct head with associated nose, billows, lobes and clefts. As the density current propagates, it generates a bore on the interface between the tail and ambient fluid. Alternatively, when the density of the surge is the same as the density of the tail, the surge moves back as a moving hydraulic jump on the interface between the tail of the flow and the ambient fluid. No intruding density current exists as there is no density difference between the bulge of fluid that accumulates at the ramp, and the tail of the forward flow into which it intrudes. In reality, when the density difference between the two fluids that accumulate at the ramp is very slight, components of both the situations described may occur.

Whether or not the density of the fluid at the ramp is equal to or greater than the density within the tail, a bore is generated on the interface between the tail and ambient fluid. This bore may take many forms between two end-member limiting cases. These are:

(i) A type A, strong bore which has a form that has many similarities with the forward moving density current (ie. asymmetry of form, billows and sometimes, lobes and clefts) (Fig. 3.1).

139

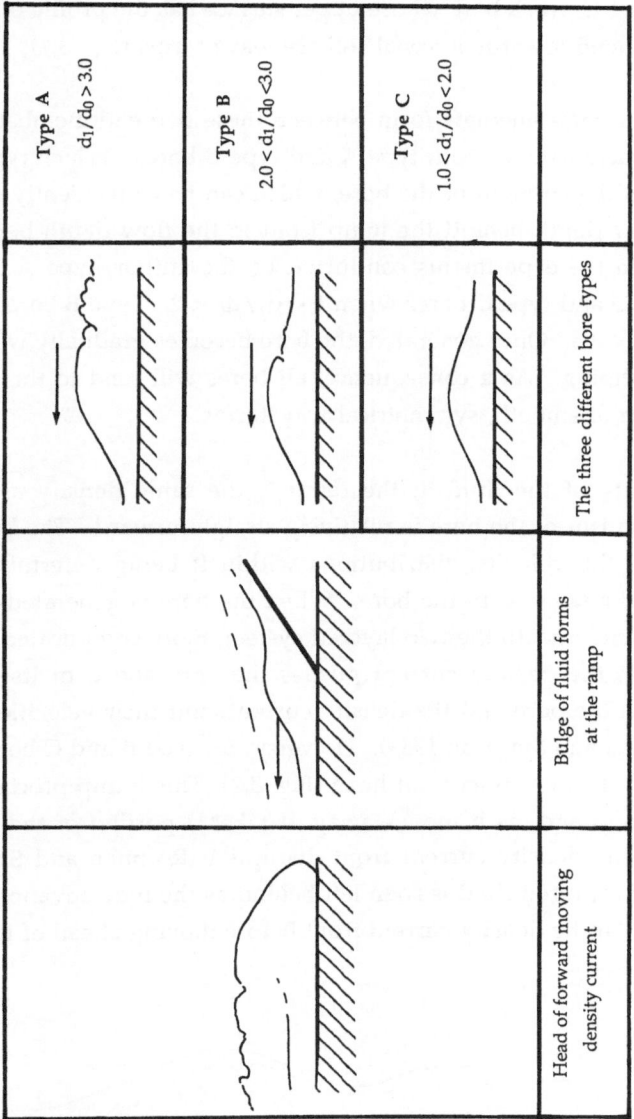

Fig. 3.1

Summary of the flows generated by the incidence of a density current upon a ramp.

(ii) A type C, weak bore whose upper surface has the profile of a group of waves. In time, the waves tend towards internal solitary wave forms (Fig. 3.1)

Bores that have an intermediate form between these two end-members (type B) have some of the flow characteristics of both type A and type B bores. Which type of bore develops is dependent upon the strength of the bore, which can be conveniently expressed as the ratio of the mean flow depth behind the jump front to the flow depth before the arrival of the bore (d_1/d_0). In the experiments conducted by the author, type A bores were generated when $d_1/d_0 > 3.5$ and type C bores when $1 < d_1/d_0 < 2$. Type B bores were found to occur when $2 < d_1/d_0 < 3.5$. Once generated, the bore becomes gradually weaker due to frictional and viscous damping. As a consequence, all bores will tend to the undular case and its' associated group of smooth, symmetrical waveforms.

When the density of the fluid at the ramp is the same density as that of the tail, the evolution mechanism of the bore is relatively straightforward. The bore advances into the layered system, the velocity distributions within it being determined by the velocities already within the tail and in the bore. When the bore is generated by the intrusion of a density current underneath the two layered system, more complicated processes occur. For bores of type A, the density current pushes the bore ahead of itself. A fixed distance remains between the bore and the density current, and their velocities are not measurably different (Rottman and Simpson 1989). However, for type B and C bores, a hump of the tail initially encloses the density current head (Fig. 3.2) This hump proceeds to move ahead of the density current and, as it moves away, it takes a portion of the density current head with it leaving the density current front disrupted (Rottman and Simpson *op cit*). This small amount of captured fluid is soon left behind as the bore advances. Subsequent waves continue to form at the density current front before moving ahead of it as just described.

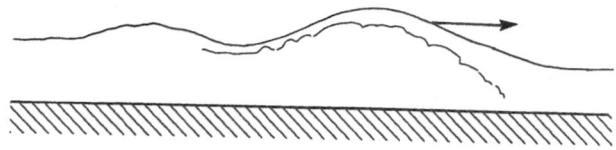

Fig. 3.2
Initial stages of the development of a bore by the intrusion of a density current under a layered system.

3.2.3 Applications of the model to more complex situations

In more complicated situations (such as the incidence of a density current upon ramps that are not orthogonal to the propagation direction or, alternatively upon obstacles within the flow that are only a fraction of the size of the current), bores will be generated that travel in a direction that opposes the propagation direction of the original current. The geometry of these unrestricted bores is determined by the geometry of the ramp or obstacle, at least for relatively simple orientations and shapes. For example, consider the incidence of a radially spreading current upon a planar ramp whose strike is at an angle to the axis of symmetry of the current. The generated bore runs parallel to the strike of the ramp and bears little relation to the original density current (Plates 2.5 to 2.7). Similarly, when a radially spreading current spreads around a hemispherical obstacle that is only a fraction of the currents width, the generated bore spreads as concentric rings, as if from a point source (Fig. 2.46).

The generation of a bore from the incidence of a density current upon a ramp or obstacle in a two-layered system is almost ubiquitous. However, if the magnitude of the obstruction (ramp angle or obstacle size), or the density of the current is not sufficient a bore will not be generated, ie. there will be no change in flow depth associated with the reverse flow. In the experimental programme conducted by the author, the only times that a bore was not generated by the incidence of a density current upon a flow obstruction was when:

(i) A density current of initial density $1.068g$ cm^{-3} was incident upon a 5° ramp in the rectangular flume tank, in a total flow depth of 5cm. The foot of the ramp was 288.5cm from the end-wall of the lock.

(ii) A density current of initial density $1.053g$ cm^{-3} was incident upon a 5° ramp, in a total flow depth of 10cm. The foot of the ramp was 270.5cm from the end-wall of the lock.

It may also be conjectured, that if a density current moves too quickly on a short ramp or over an obstacle, then the flow may not decelerate significantly; no bulge of fluid will form and consequently a bore, or any significant reverse motion, will not occur. However, when currents with a smaller velocity impinge upon this same obstruction a bore may form with ease.

3.2.4 Velocity characteristics of the generated bores

Initially, in relation to the situation in which the density of fluid in the bulge is the same as the density of the lower layer, the velocity distributions of the two end-member bore types will be different. For type A bores, the near-bed velocity profile at a point will be similar to that of the original forward density current (Fig. 2.3). An initial, almost instantaneous velocity increase will occur as the bore arrives at the point of measurement. This will be followed by flow deceleration. This conclusion is not unexpected as the density current of the forward flow is simply a bore of maximum strength, ie. d_1/d_0 is equal to infinity. On the other hand, a pulsating velocity profile will be associated with the passage of a weak bore (Fig. 2.21). As each wave passes, there will be an increase in velocity to a maximum as the wave crest passes the point of measurement, followed in turn, by a decrease of velocity. Between each wave the velocity is much lower, although it is still in the direction of bore propagation. As the waves become more solitary, the distance between individual waves increases and the velocity in the flow between the waves tends towards zero. It is important to note that in time, all bore types will weaken. Once the flow is a type C bore it continues to weaken and the form of the profile progressively approaches that of a group of internal solitary waves. Therefore, during the flow of a bore generated by the incidence of a density current upon a flow obstruction, it may pass from a strong type A bore to a very weak type C bore (Fig. 2.35 and 2.36). As a result, the velocity characteristics of the generated bore would gradually change from those typical to the forward flow of the original density current to those typical of the passage of a group of internal solitary waves.

3.3   The model with respect to turbidity current systems

The model presented so far is suitable for the description of density current phenomena and will probably be adequate to account for the incidence of very fine-grained, well sorted turbidity currents developed on obstructions encountered during flow. It is important to establish whether these laboratory phenomena could occur in real-scale turbidity currents and if so, what the probable magnitude of the bore's features such as wavelength, would be. Consequently, if they are generated in natural turbidity currents, it is desirable to know: (i) what sort of turbidity currents are favoured (eg. grain-size distribution, sorting etc.) and, more importantly, (ii) whether the bore will generate any diagnostic sedimentary structures which will enable identification in the sedimentary rock record.

To enable turbidity currents to generate bores, the most crucial factor is the presence of a basin geometry that will allow the run-up of turbidity currents. Such a geometry could be supplied by the basin margins. Alternatively, intrabasinal highs, such as fault scarps or seamounts, could also provide suitable obstructions to the flow of turbidity currents. The current must then have enough momentum to travel up the obstruction and whilst sediment falls out of suspension due to deceleration, enough must remain in suspension to generate the bore. A current that will fit these criteria will probably have to have a high concentration of fine-grained sediment, as the low settling velocity of mud and fine-grained sediment will enable the density difference between the current and the ambient sea-water to be maintained for a long time. In addition, a high concentration of fine-grained sediment will mean that not all the sediment in suspension will fall out as the current decelerates whilst moving up the obstacle.

Let us consider a hypothetical, moderately well-sorted muddy turbidity current that impinges upon a planar basin margin that has a slope of approximately 20°. The flow decelerates as it moves up the slope and the coarser sediment rapidly settles out. A bulge of suspended sediment forms near the base of the basin margin which soon begins to propagate back through, or under the tail of the still forward moving turbidity current. There will be sufficient density excess in suspension to form the bulge, as the processes occurring at the foot of the ramp will take a fraction of the time relative to that taken for the finer-grained sediment to fall out of suspension. As shown by Parker *et al* (1987), fine-grained turbidity currents can entrain coarser-grained material. It is possible that the collapsing head higher up the margin may re-entrain some of the coarser sediment that has only just settled out from suspension. In addition, as the bulge propagates through the tail (which has a smaller flow power), it may also re-entrain sediment that had been recently deposited.

As the bore propagates it will be continually damped by friction and if the flow is on a slight positive gradient, it will also be slowed down by gravity. As suggested by the model, the nature of the bore will be governed by the ratio d_1/d_0. The actual values of d_1/d_0 that determine whether a type A, B or C bore will form will probably be different than in the model as the effect of sediment fall-out will: (i) suppress turbulence and therefore, may mean that type A bores will form at higher d_1/d_0 values than in simple density currents and (ii) cause the density of the flow to decrease more rapidly hence affecting the bore strength. However, all three bore types should be possible.

The model suggests that the bore evolution mechanisms that occur will depend upon the density of the fluid that accumulates at the bulge relative to the density of the tail. Although, both types of bore generation could occur, it is unlikely that the density of the bulge would be the same as that of the tail and consequently, the fluid of the bulge probably flows under the tail so as to generate a bore on the interface above. There is also the possibility that in very sandy currents, so much sediment is deposited at the foot of the basin margin that the residual current is less dense than the tail. In this circumstance, the bulge would intrude along the horizon that was the same density as either an intrusive density current (Britter and Simpson 1981) or a plume. Sediment from either of these features would fall out from suspension as a 'sediment rain'.

As the bore passes, the velocity should be great enough in both the strong bore and the waves of a weak bore to affect recently deposited sediment. The sediment, not being lithified, may not withstand erosive events. Although sediment may initially be re-entrained, as the bore begins to decelerate sediment will be deposited and may result in the development of tractional bed-forms. As a strong bore passes, the potential sequence of sedimentary structures will reflect deposition from a waning flow and hence, may have many Bouma-like characteristics. Much of the sediment that is deposited will have been transported by the bore from locations that are nearer to the site of the flow obstruction. As a weak bore passes, formation of bed-forms is most likely during the passage of the associated waves. Between the waves, suspension fall-out may be resumed and drape any bedforms produced. So, the passage of a group of waves may generate bedforms separated by drapes of finer-grained material. Overall, subsequent waves will become weaker, transport finer-grained sediment and generate bedforms associated with a lower flow power and therefore smaller velocity.

The propagation direction of the generated bore is governed primarily by the geometry of the obstruction. As a result, the flow generated by the incidence of a turbidity current on for example, a slope whose strike is oblique to the propagation direction of the original turbidity current, will not necessarily retrace the path of the forward flow. If the current did not retrace the original path then, upon intruding into suspension free water it must become a turbidity current *per se*, and the sedimentary structures produced will reflect this.

In theory, therefore, the propagation of a weak bore could produce a characteristic sedimentary sequence. The control on whether these structures will actually occur and if so, how significant a part of the turbidite they will be is predominantly determined by the wavelength of the waves of the bore. The consideration of natural turbidity currents from which some data have been measured in conjunction with the experimental results described in chapter 3 should enable wavelengths to be estimated. Possible calculated wavelengths and the consequences and problems encountered in deriving them are outlined in the following section.

3.4   Bores and recent turbidity currents.

A detailed knowledge of real currents is required in order to predict the type of bore and its dimensions that would be generated in natural turbidity currents. Unfortunately, there has been only limited success in the acquisition of data from natural turbidity currents and the knowledge of deep marine turbidity currents remains very limited. A summary of some of the data acquired in studies of natural, recent turbidity currents is shown in Table 3.1.

Flow velocities have been measured from cable breaks (eg. Heezen and Ewing 1952) and, to varying degrees of success, with current meters (eg. Prior *et al* 1987). In addition, flow velocities have been calculated using grain-size distributions, channel morphology (eg. Bowen *et al* 1984) and levee heights (Komar 1969). Flow thicknesses have not been

Location	Primary data source	Site type	Velocity (m s-1)	Flow thickness (m)	Slope	Grain size (mm)	Source
Bute Inlet, British Columbia	Current meters	Channel	0.06	<13	?	Median 0.05	Hay et al (1982)
"	"	"	3.35 0.75	<30	0.71° 0.58°	Coarse sand to mud	Prior et al (1987)
Laurentian Fan	Cable breaks	Continental rise	25.7		<1°		Heezen and Ewing (1952)
"	"	Oceanic floor	6.2		"		
	"	Upper fan	18	<300		Maximum = 20	Piper et al (1985)
Monterey Fan	Levee heights	Channel	4 - 20	100	0.08° - 0.5°	0.1 - 0.2	Komar (1969)
Navy fan	Grain size & cross flow slope	Basin	0.12	75	1:500		Bowen et al (1984)
		Channel	0.75	<70	1:60		
			0.73	20	1:100		
		Mid-fan	0.32	15	1:50		
Scripps canyon	Current meters	Channel	1.9	3 - 10	1:12.5	0.15	Inman et al (1976)

Table 3.1: Data derived from studies of natural turbidity currents.

measured accurately, often being inferred from channel depths and vegetation distribution (eg. Inman *et al* 1976). Finally, data for the density distribution within turbidity currents has not been measured directly and has only been estimated by concentration calculations from grain-size distributions (eg. Bowen *et al* 1984).

With such limited knowledge of the flow characteristics of natural turbidity currents extrapolation of the laboratory results from the present work can only be tentative. Indeed, comparison of small-scale, salt solution density currents with large scale, sediment suspension flows may be an un-realistic step to take. However, it may provide some initial ideas and parameters which when examined, may lead forward to more detailed work for the future.

The most commonly measured properties of natural turbidity currents are the flow velocity and the flow thickness. Assuming that the measured value of the flow velocity is equivalent to U_h, as measured in the experiments, and that the measured value of the flow thickness is equivalent to d_0, then L/d_0 and U_b/U_h as measured in the experiments can be used to derive values for natural turbidity currents. Such considerations generate values of L for a weak undular bore of nearly 2km that would be associated with velocities of 18 to 22m s^{-1} for the Grand Banks turbidity current. As L is not a measure of the total effective length of the wave, the length over which the wave may be active is likely to be considerably longer. Thus in this instance, the wave could take considerably more than the calculated maximum duration of 106sec. If the same turbidity current generated a strong bore which resembled a density current, then it would have a maximum 'head height' of 750m with velocities that vary between 13 and 21m s^{-1}.

Although such calculations generate geologically reasonable results, several of the assumptions must be discussed before accepting that these numbers are significant. The assumptions that (i) the ratio L/d_0 is a valid one to use, (ii) that d_0 can be assumed to equal the measured flow thickness and that (iii) the laboratory ratios can immediately be applied by scaling up to natural turbidity currents are fundamental in the calculation of L for natural conditions. In fact, d_0 will probably be considerably less than the measured flow thickness as the latter is probably a measure of the maximum thickness of the body whilst d_0 is most likely to be the thickness of the tail of the forward current. In laboratory experiments that use salt solutions, d_0 is controlled by the tank geometry and by the controlled, constant volumes of the fluids in the experiments. This means that in the experiments, by the time the bore is formed the value of d_0 will be constant throughout the tank. This is probably not the case in the natural environment. So, although in the experiments L/d_0 remains approximately constant with time (only being gradually changed by frictional dampening), a variation in this ratio with time is likely to occur in natural systems.

The most important restriction upon the extrapolations made is that it assumes that a sediment suspension acts in the same way as a salt solution density current. Experiments conducted by the author indicated that although the same features are generated by sedimentary suspensions, modifications of the laboratory conditions were required in order to compare flows that originally had the same density. For example, the first waveform of the bore generated by a sediment suspension current had a velocity that was 54% of the equivalent saline solution flow, although the amplitudes were similar (see section 2.10). The waves generated by the turbidity current were asymmetrical and unevenly spaced, whereas for the equivalent saline solution experiment they were approximately symmetrical and evenly distributed.

The present experimental work provides a good qualitative description of the flow processes that are associated with the incidence of a density current upon a flow obstruction. However, use of the experiments to directly obtain values of the magnitudes of the flow phenomena of the bore (eg. wavelength) must be treated with caution. This is because sediment suspension currents require some modification of the results determined during the present experimental programme.

Considerations of the potential magnitude of the type C bore wavelengths for the Grand Banks turbidity current indicates that they could generate features that would be preserved within the sedimentary record. This is discussed further in the following sections.

3.5 Bores in the Windermere Group, England

The following is a extract from Kneller *et al* (1991), providing further evidence of another sedimentary sequence in which the generation of bores has played an important part.

"We have studied a thick and laterally extensive turbidite sequence from the Windermere Group of northwest England (Fig. 3.3). This succession consists largely of late Wenlock to Ludlow marine siliciclastic turbidites deposited in a peripheral foreland basin. Sole structures on the sandstone turbidites show diverse palaeocurrent orientations, but the majority indicate flow parallel to the local basin axis (Fig. 3.3), in part reflecting basement trends. Ripples commonly show palaeoflow approximately transverse to the local basin orientation. We have examined palaeocurrent data from a 40km long segment of the basin oriented northeast/southwest (our data and Norman 1961). Sole structure flow directions (S) are predominantly axial, and mostly to the southwest. Bedding surfaces with asymmetrical ripples are common, as are bedding-normal surfaces with ripple cross-lamination. The degree of ripple asymmetry is commonly slight. The ripples form part of the accretionary sequence of structures within the normally graded deposits of relatively

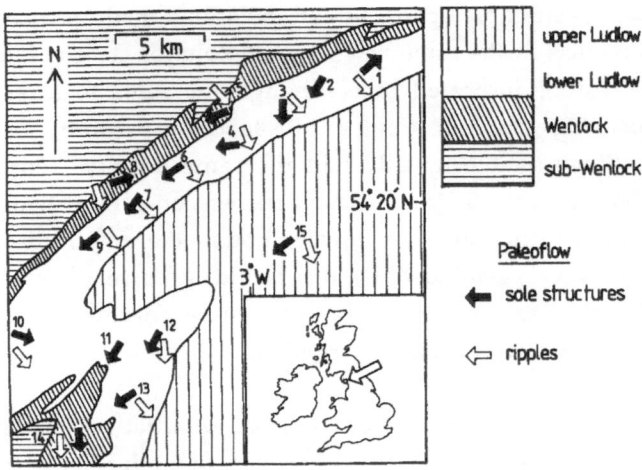

Fig. 3.3

Outline geological map of Cumbria, northwest England (inset shows area of main map), showing palaeocurrent vector means of Wenlock and Ludlow age sands turbidites within the Windermere Group.

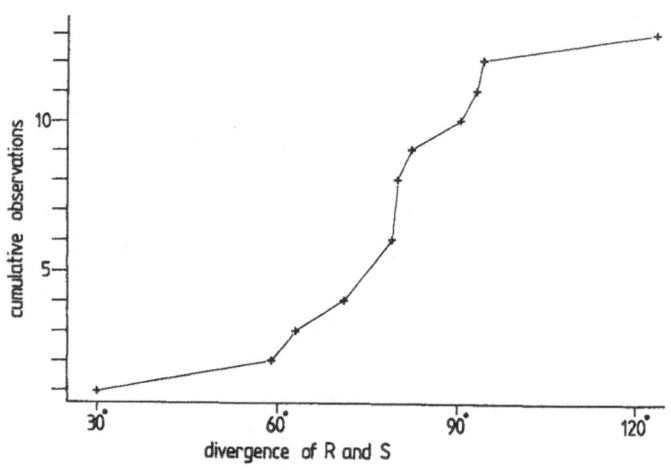

Fig. 3.4

Cumulative plot of angular differences between S and R directions measured on the same bed, where the discrepancy is 30° or greater; from Wenlock and Ludlow sand turbidites of the Windermere Group.

low-density turbidity currents, and they are also present in the upper graded parts of massive sandstone deposits of high-density flows (Lowe 1982). The cross-lamination often develops progressively from underlying parallel lamination, with no sign of a break in sedimentation; each bed, therefore, forms the deposit of a single flow event. Ripple flow directions (R) are almost universally transverse, predominantly towards the southeast (Fig. 3.3). The distributions of R and S directions can be distinguished with better than 95% confidence (Mardia 1972).

High angles between R and S directions occur within individual beds (Fig. 3.3). Turbidites with axial S directions show divergence of R directions to right or left, at angles ranging from 60° to 127°. By contrast, beds with sole structures indicating transverse currents generally show agreement between R and S directions. This pattern of near perpendicular R and S directions persists through the sequence of sandstone turbidites (Fig. 3.4)."

"Any explanation must account for the apparent continuity of deposition within beds, the lack of consistency in sense of rotation between S and R, and the scarcity of intermediate R directions. Reworking of the tops by clear-water bottom currents can be rejected because the ripples are clearly part of a sequence of structures related to waning flow; this is not to dispute the occurrence of such re-worked facies elsewhere (Hsu 1964). Erosion of the sole structures by non-depositing axial currents and their subsequent filling by deposits of transverse flows is also unlikely; many of these sandstones are the deposits of comparatively high density turbidity currents, and it is implausible that such flows would have produced no erosional effects; crossing sets of sole structures on the same bed are occasionally present, but the obliquity is seldom more than 20°. Amalgamation of the deposits of simultaneous lateral and axial flows fails to account for either the consistent angular relation between the R and S directions or the petrographic similarity of sandstones throughout individual beds. Deflection of downslope currents (S) into a contour following mode (R) at the base of the slope is in contradiction of the basin geometry. Deflection by the Coriolis force fails to explain deflections in the wrong sense, or the virtual absence of intermediate current directions. Banking of flows, with a progressive change from momentum-driven along-slope flow (S) to down-slope gravitational flow (R) (Seilacher and Meischner 1965; Prentice 1960; Woodcock 1976), is also inconsistent with the scarcity of intermediate R directions, and requires exceptional basin configurations."

"We suggest that the ripples were generated by reflections from topographic features oblique to the incident flow.........Applying these observations* to turbidity currents in nature, we envisage the generation of large scale 'reflections' within constricted basins, both from their margins and from intrabasinal topography associated with volcanic edifices, normal faulting, and thrusting. Where large turbidity currents are introduced into

---

*Refers to experiments in chapter 2.9

elongate basins, much of the flow disperses axially along the basin floor, eroding beneath the head and depositing much of the sand grade suspended load from the body of the flow. Sand-filled erosional features produce casts indicating palaeoflow (S direction) parallel to the basin axis. Internal solitary waves are generated where radial spreading is constrained by topography. They flow back from the reflecting surface, perpendicular to its strike, and sweep across the residual body of the turbidity current. Under some circumstances fluid translation within the waves may re-suspend recently deposited fine sand from the bed or rework it as bedload. Observed ripple geometries are consistent with formation by unidirectional flow with a pulsatory component. For most basin configurations, the direction of wave propagation will be at a high angle to the original flow; resultant bedforms such as ripples will indicate flow (R directions) at a high angle to S directions. Because R directions (the direction of propagation of the waves) are perpendicular to the strike of the reflecting surface, they may provide evidence of the orientation of ancient basin margins. In basins with asymmetric profiles, preferential reflection might be expected from the margin with the steeper slope.

We suggest that the enigmatic palaeocurrent patterns of the Windermere Group, and other turbidite sequences, can be explained by internal solitary wave generation at the relief associated with basin margins."

3.6 CONCLUDING REMARKS

This thesis provides both a qualitative and quantitative study of the consequences of the incidence of both restricted and unrestricted density currents upon various low-angled ramps and obstacles. The study shows that bores will be generated which may have velocity characteristics similar to the forward flow. For the case of restricted density currents, the control of variables such as the angle of the ramp, the thickness of the generated flow, velocity of the original flow and the density difference between the original flow and the generated bore have been considered. This has enabled a model to be determined for the straight forward case of the incidence of density currents (where the density excess is due to a denser fluid rather than a sediment suspension) upon ramps and obstacles.

Although this study has proved that the same features can be produced by the incidence of a turbidity current upon a ramp, it has also showed that modifications are required to the salt solution model before it can be applied, in a quantitative fashion, to sediment suspensions. Without knowledge of these modifications it is difficult to apply this salt-solution model, quantitatively, to suspension currents in the laboratory. However, qualitatively, it should provide a useful tool. It is therefore, not surprising that application of the aforementioned model to natural turbidity currents can only, at this stage, be qualitative.

One problem with the application of laboratory models of turbidity current flow to natural turbidites, is that there still remains a large gap in our understanding of the natural phenomenon. Data that have been acquired are often unreliable and do not provide a complete picture of the physical properties of turbidity current flow. In addition, measurements are usually made in a canyon or channel, whereas most sediment deposition and interaction of the flow with obstructions will occur on submarine fans and the basin floor. Thus, the data that is available for natural turbidity currents is not necessarily of use when it is often details of the flow outside of the canyon that are required.

Evidence to support the model presented in this thesis can be proposed from the sedimentary record. As illustrated in both the Cloridorme Formation and the Windermere Group, the generation of bores could provide a suitable explanation for at least some of the unusual sedimentary sequences that occur in the Formation. Again, knowledge must remain qualitative on the whole. Quantitative data are usually derived by calculation of shear stresses from either bed-form transitions or grain-size analyses. These methods use data derived from the study of steady flows that are in an equilibrium state in which sediment is initially stationary on the bed, subsequently being entrained by the stream passing above. A relationship would therefore be expected between the maximum grain size entrained and the flow power of the stream. In turbidity currents however, flow only occurs because there is sediment in suspension to begin with and therefore the relationship between flow power and grain size transported (if it exists at all) could be very complex.

In conclusion, further advances within turbidity current research and in particular, continuation of this present study, needs to be based upon a solid data-base of (i) the physical properties of natural turbidity currents and (ii) an in depth knowledge of what physical parameters govern bedform formation including the influence of factors, such as sediment load fall out. The latter will only be achieved by detailed experimental programs in apparatus (not lock-exchange) that is specifically designed for the study of the development of sedimentary structures in turbidity currents.

As regards the expansion of the study presented in this thesis in the future several aspects could be pursued:

(i) A **detailed** study of effects of the incidence of restricted and unrestricted density currents on ramps and obstacles, generating similar relationships as this study has for the case of restricted currents being incident on low-angled ramps,

(ii) Consideration of the incidence of density currents upon irregular topography and a variety of obstacle shapes,

(iii) Determination of the different consequences of bores generated: (a) by the intrusion of a denser current into a two layered system and (b) those generated by the release of a dense fluid into a layered system (where the basal layer has the same density as the released fluid), upon near-bed velocities and sediment transport.

Perhaps most important however, will be to adapt the salt solution model to one which can be applied in a quantitative fashion to both natural turbidity currents and also the sedimentary record. As regards the sedimentary record it would be advantageous if more examples of turbidites with evidence for reversed flow, such as the Cloridorme Formation, could be found as it will add substance to the model presented in this thesis, perhaps of a more substantial nature than consideration of natural turbidity currents can provide at present.

REFERENCES

Akiyama, J. and Stefan, H. G., 1987. Onset of underflow in slightly diverging channels. J. Hydraulic Eng. **113**: 825 - 845.

Akiyama, J. and Stefan, H. G., 1988. Turbidity current simulation in a diverging channel. J. Water Resources. **24**.4: 579-588.

Allen, J. R. L, 1970. Sequence of sedimentary structures in turbidites, with special reference to dunes. Scott. J. Geol. **6**: 146-161.

Allen, J. R. L., 1971. Mixing at turbidity current heads and its geological applications. J. Sed. Pet. **41**: 97-113.

Allen, J. R. L., 1982. Sedimentary Structures: Their character and physical basis. Developments in Sedimentology, **30A, 30B**. Elsevier, Amsterdam. pp.593 and 663 respectively.

Allen, J. R. L, 1985. Principles of Physical Sedimentology. George Allen and Unwin Ltd., London, pp 272.

Bagnold, R. A., 1962. Autosuspension of transported sediment; turbidity currents. Proc. Roy. Soc. London. **A265**: 315-319.

Bagnold, R. A., 1966. An approach to the sediment transport problem from general physics. U.S. Geol. Survey Prof. Pap. **422-I**.

Baines, P. G., 1984. A unified description of two-layer flow over topography. J. Fluid Mech. **146**: 127-167.

Baines, P. G., 1987. Upstream blocking and airflow over mountains. Ann. Rev. Fluid Mech. **19**: 75-98.

Baines P. G. and Davies, P. A., 1980. Laboratory studies of topographic effects in rotating and/or stratified fluids. In, Orographic Effects in Planetary Flows, GARP Publ. No. 23, 233-299.

Banerjee, I., 1977. Experimental study of the effect of deceleration on the vertical sequence of sedimentary structures in silty sediments. J Sed. Pet. **47**: 771-783.

Barr, D. I. H., 1963. Densimetric exchange flow in rectangular channels. I: Definitions, review and relevance to model design. La Houille Blanche, **7**: 739-766.

Beeden, D. R., 1983. Sedimentology of some turbidites and related rocks from the Cloridorme Group, Ordovician, Quebec. MSc. Thesis, (Unpubl.). McMaster Univ., Canada.

Benjamin, T. B., 1966. Internal waves of finite amplitude and permanent form. J. Fluid Mech. **25**: 241 - 270.

Benjamin, T. B., 1967. Internal waves of permanent form in fluids of great depth. J. Fluid Mech. **29** : 559 - 592.

Benjamin, T. B., 1968. Gravity currents and related phenomena. J. Fluid Mech. **31**: 209-248.

Benjamin, T. B. and Lighthill, M. J., 1954. On cnoidal waves and bores. Proc. Roy. Soc. Lond., **A224**: 448-460.

Bennetts, K. R. W. and Pilkey, O. H., 1976. Characteristics of three turbidites, Hispaniola-Caicos Basin. Bull. Geol. Soc. Am. **87** : 1291 - 1300.

Benney, C. J., 1966. Long non-linear waves in fluid flows. J. Math. Phys. **45** : 52 - 54.

Bouma, A. H., 1962. Sedimentology of some flysch deposits. Elsevier, Amsterdam., pp. 167.

Bowen, A. J., Normark, W. R. and Piper, D. J. W., 1984. Modelling of turbidity currents on the Navy Fan, California Continental Borderland. Sedimentology. **31**: 169-185.

Bradshaw, P., 1971. An introduction to turbulence and its measurement. 1st Ed. , Aeronautic Dept., Imperial College. Permagon Press, New York. pp228.

Britter, R. E. and Simpson, J. E., 1978. Experiments on the dynamics of a gravity current head. J. Fluid Mech. **88**: 233-240.

Britter, R. E. and Linden, P. F., 1980. Motion of the front of a gravity current travelling down an incline. J. Fluid Mech. **99**: 531-543.

Buffington, E. C., 1961. Experimental turbidity currents on the sea floor. Bull. AAPG. **45**: 1392-1400.

Chu, V. H. and Baddour, R. E., 1977. Surges, waves and mixing in two layer density stratified flow. In, Proc. 17th. Congr. Intl. Assn. Hydraulic Res., Vol 1, 303-310.

Davis, R. E. and Acrivos, A., 1967. Solitary waves in deep water. J. Fluid Mech. **29**: 593 - 607.

Ellis, D., 1982. Palaeohydraulics and computer simulation of turbidites in the Marnoso-Arenacea, Northern Appenines, Italy. PhD Thesis. (Unpubl.) Univ. of St. Andrews, Scotland.

Ellison, T. H. and Turner, J. S., 1959. Turbulent entrainment in stratified flows. J. Fluid Mech. **6**: 423-488.

Enos, P., 1969 Cloridorme Formation, Middle Ordovician Flysch, Northern Gaspe Peninsula, Quebec. Geol. Soc. Am. Special Paper **117**. pp.66.

Farmer, D. M., 1978. Observations of long non-linear internal waves in a lake. J. Phys. Oceanography. **8**: 63-73.

Fisher, R., 1971. Features of coarse grained high concentration fluids and their deposits. J. Sed. Pet. **41**: 916-927.

Garcia, M. and Parker, G., 1988. Entrainment of bed sediment by density underflows. Proc. ASCE Hydraulics Conf., Colarado Springs.

Georgeson, E. H. M. 1942. The free streaming of gases in sloping galleries. Proc. Roy. Soc. Lond. **A180**: 484-493.

Ghosh, J. K., Mazumder, M. R. Saha. and Sengupta, S., 1986. Deposition of sand by suspension currents: Experimental and theoretical studies. J. Sed. Pet. **56**: 57-66

Gould, H. R., 1951. Some quantitative aspects of Lake Mead turbidity currents. SEPM Spec. Publ. No. 2., 34-52.

Grimshaw, R. H. J., 1978. Long non-linear internal waves in channels of arbitrary cross-section. J. Fluid Mech. **86**: 415 - 431.

Grimshaw, R. H. J., 1983. Solitary Waves in Density Stratified Fluids. In, U. Nigul and J. Engelbrecht (eds.), Nonlinear Deformation Waves. Springer, Berlin. pp.431-447.

Guy, H. P. , Simons, D. B. and Richardson, E. V., 1966. Summary of alluvial channel data from flume experiments , 1956-61. Prof. Paper U.S. Geol. Surv. **462-1**: 1-96.

Hand, B. M., 1974. Supercritical flow in density currents. J. Sed. Pet. **44**: 637-648.

Hand, B. M., Middleton, G. V. and Skipper, K., 1972. Antidune cross-stratification in a turbidite sequence, Cloridorme Formation, Quebec. Discussion. Sedimentology **18**: 135 - 138.

Hay, A. E., Burling, R. W. and Murry, J. W., 1982. Remote acoustic detection of a turbidity current surge. Science **217**: 833-835.

Hersey, J. B., 1965. Sediment ponding in the deep sea. Bull. Geol. Soc. Am. **76**: 1251 - 1260.

Heezen, B. C. and Ewing, M., 1952. Turbidity currents and submarine slumps, and the 1929 Grand Banks earthquake. Am. J. Sci. **250**: 849-873.

Hersey, J.B., 1965. Sediment ponding in the deep sea. Bull. Geol. Soc. Am. **76**: 1251-1260.

Hesse, R., 1982. Cloridorme Formation. In:- Palaeozoic continental margin sedimentation in the Quebec Appalations. (Eds. Hesse, Middleton and Rust), XIth. Int. Congr. Sedtly.,McMaster Univ., Hamilton, Canada. Field Trip Guide Book, Excursion 7B. 126-138.

Kuenen, Ph. H. and Hubert, K.L, 1964. Bibliography of turbidity currents, 222-246., In, Developments in Sedimentology No.3, A.H.Bouma and A. Brower (eds.), Elsevier, Amsterdam, 264pp

Kuenen, Ph. H. and Migliorini, C.I., 1950. Turbidity currents as a cause of graded bedding. J. Geol. 58: 91-127.

Laval, A., Cremer, M., Beghin, P. and Ravenne, C., 1988. Density surges: Two-dimensional experiments. Sedimentology 35: 85-103.

Le Blond, P. and Mysak, L. A., 1978. Waves in the Ocean. Elsevier Oceanography Series No. 20., Elsevier, Amsterdam, 602pp.

Long, R. R., 1956. Solitary waves in one and two fluid systems. Tellus 8: 460 - 471.

Long, R. R., 1965. On the Boussinesq approximation and its role in theory of internal waves. Tellus 17: 46 - 52.

Long, R. R., 1970. Blocking effects in flow over obstacles. Tellus, 22: 471-480.

Lowe, D. R., 1982. Sediment gravity flows: II. Depositional models with specific reference to deposits of high density turbidity currents. J. Sed. Pet., 52: 279-297.

Lowe, D. R., 1988. Suspended load fallout rate as an independent variable in the analysis of current structures. Sedimentology. 35: 765-776.

Luthi, S., 1980. Some new aspects of two-dimensional turbidity currents. Sedimentology. 28: 97-105.

Marjanac, T., 1990. Reflected sedimentary gravity flows and their deposits in the flysch of Middle Dalmatia, Yugoslavia. Sedimentology 37: 921-930.

Mardia, K. V., 1972. Statistics of directional data. London Academic Press. pp357.

Maxworthy, T., 1980. On the formation of non-linear internal waves from the gravitational collapse of mixed regions in two and three dimensions. J. Fluid Mech. 96 : 4 - 64.

Meiss, J. D. and Pereira, N. R., 1978. Internal wave solitons. Phys. Fluids 21: 700 -702.

Middleton, G. V., 1966a. Small scale models of turbidity currents and the criterion for autosuspension. J. Sed. Pet. 36: 202-208.

Middleton, G. V., 1966b. Experiments on density and turbidity currents. I. The motion of the head. Can. J. Earth Sci. 3: 523-545.

Middleton, G. V., 1966c. II. Uniform flow of density currents. Can. J. Earth Sci. 3: 627-645.

Middleton, G. V., 1967. Experiments on density and turbidity currents. III. Deposition of sediment. Can. J. Earth Sci. 4: 475-505.

Middleton, G. V., 1976. Hydraulic interpretation of sand size distributions. J. Geol. 84: 405-426.

Middleton, G. V. and Hampton, M. A., 1973. Sediment gravity flows: Mechanics of flow and deposition. In, G.V. Middleton and A.H. Bouma (eds.) Short Course Notes, Pacific Sect. Soc. Econ. Palaeont. Mineral., 1-38.

Middleton, G. V. and Hampton, M. A., 1976. Subaqueous sediment transport and deposition of sediment gravity flows. In, D.J. Stanley and D.J.P. Swift (eds.), Marine Sediment Transport and Environmental Management. Wiley, New York. pp 197-218.

Middleton, G. V. and Neal, W. J., 1989. Experiments on the thickness of beds J. Sed. Pet. 59: 297-307.

Miles, J.W., 1980. Solitary waves. Ann. Rev. Fluid Mech. 12 : 11 - 43.

Muck, M.T. and Underwood, M.B., 1990. Upslope flow of turbidity currents: A comparison among field observations, theory and laboratory models. Geology 18: 54-57.

Mysak, L.A., 1984. Non-linear internal waves. in, Hydrodynamics of Lakes: 129 - 152. K. Hutter. (ed.). Int. Centre for Mech. Sci; courses and lectures No. 286. Springer-Verlag New York. pp. 129-152.

Natland, M. L. and Kuenen, Ph. H., 1957. Sedimentary history of the Ventura Basin California, and the action of turbidity currents. SEPM. Spec. Publ. No. **2**: 76-107.

Norman, T. N., 1961. The geology of Silurian strata in the Blawith area, Furness: (unpub. PhD thesis), University of Birmingham, England. pp294.

Ono, H., 1975. Algebraic solitary waves in statified fluids. J. Phys. Soc. Japan **39** :1082 - 1091.

Ostrovsky, L. A. and Stepanyants, Yu. A., 1989. Do internal solitons exist in the ocean ? Rev. Geophys. **27** : 293 - 310.

Pantin, H. M., 1979. Interaction between velocity and effective density in a turbidity flow; Phase-plane analysis, with criteria for autosuspension. Marine Geology. **31**. 59-99.

Pantin, H. M. and Leeder, M. R., 1987. Reverse flow in turbidity currents: the role of internal solitons. Sedimentology **34** : 1143 - 1155.

Parea, G. C. and Ricci Lucci, F., 1975. Turbidite key-beds as indicators of ancient deep sea plains. Proc. IX Int. Congr. Sed., Nice, Theme **1** : 235 - 245.

Parker, G., Fukushima, Y. and Pantin, H. M., 1986. Self accelerating tubidity currents. J. Fluid Mech. **171**: 145-181.

Parker, G. and Fukushima, Y., 1987. Experiments on turbidity currents over an erodible bed. J. Hyd. Res. 25, No.1.: 123-147.

Pickering, K. T. and Hiscott, R. N., 1985. Contained (reflected) turbidity currents from the Mid-Ordovician Cloridorme Formation, Quebec, Canada : an alternative to the antidune hypothesis. Sedimentology **32** : 373 - 394.

Pickering, K. T., Hiscott, R. N. and Hein, F. J., 1989. Deep marine Environments. Unwin and Hyman, London, 416pp.

Piper, D. J. W. 1978. Turbidites, muds and silts on deep sea fans and abyssal plains. In, D.J.Stanley and G. Kelling (eds.). Sedimentation in Submarine Canyons, Fans and Trenches. Hutchinson and Ross, pp163-176.

Piper, D. J. W., Shor, A. N., Farre, J. A., O'Connell, S. and Jacobi, R., 1985. Sediment slides and turbidity currents on the Laurentian Fan : Sidescan sonar investigations near the epicentre of the 1929 Grand Banks earthquake. Geology **13**: 538-541.

Prentice, J. E., 1960. Flow structures in sedimentary rocks: J. Geol. **68**: 217-225.

Proir, D. B., Bornhold, B. D., Wiseman, W. J. and Lowe, D. R., .1987. Turbidity current activity in a British Columbia fjord. Science **237**: 1330-1333.

Ricci-Lucci, F. and Valmori, E., 1980. Basinwide turbidites in a Miocene oversupplied deep sea plain: a geometric analysis. Sedimentology **27** : 241-270.

Riddel, J. F., 1969. A laboratory study of suspension effect density currents. Can. J. Earth Sci. **6**: 231-246.

Rottman, J. W. and Simpson, J. E., 1983. Gravity currents produced by instantaneous releases of a heavy fluid in a rectangular channel. J. Fluid Mech. **135**: 95-110.

Rottman, J. W. and Simpson, J. E., 1989. The formation of internal bores in the atmosphere: A laboratory model. Q. J. R. Meteorol. Soc.115: 941-963.

Russel, J. S., 1838. Report of the committee on waves. Report Meeting Brit. Assoc. Adv. Sci. (Liverpool 1837). John Murray, London.

Ryan, W. B. F., Fifield, W. and Hersey, J. B., 1965. Sediments on the Tyrrhenian Abyssal Plain. Bull. geol. Soc. Am. **76** : 1261 - 1282.

Schmidt, W., 1911. Zur mechanick der Boen. Z. Meteeorol. **28**: 355-362.

Segur, H. and Hammack, J. L., 1982. Soliton models of long internal waves. J. Fluid Mech. **118** : 285 - 304.

Seilacher, A. and Meischer, D., 1965. Fazies-analyse im palazoikum des Oslo-Gebeites. Geolische Rundschau. **54**: 596-619.

Siegenthaler, C., Hsu, K. J. and Kleboth, P., 1984. Longitudinal transport of turbidity currents - A model study of Horgen events. Sedimentology **31**: 187-193.

Simons, D. B. and Albertson, M. L, 1961. Flume studies using medium sand (0.45mm). Water Supply Paper. U.S. Geol. Surv. No. **1498-A.**, 76pp.

Simpson, J. E., 1969. Comparison between laboratory and atmospheric density currents. Q. J. Roy. Met. Soc. **95**: 758-765.

Simpson, J. E., 1972. Effects of the lower boundary on the head of a gravity current. J. Fluid Mech. **53**: 759-768.

Simpson, J. E., 1982. Gravity currents in the laboratory, atmosphere and ocean. Ann. Rev. Fluid Mech. **14**: 213-234

Simpson, J. E., 1987. Gravity currents: In the Environment and the Laboratory. Ellis Horwood Series in Environmental Science. John Wiley and Sons Ltd., Chichester. pp 244.

Simpson, J. E. and Britter, R. E., 1979. The dynamics of the head of a gravity current advancing over a horizontal surface. J. Fluid Mech. **94**: 477-495.

Skipper, K., 1971. Antidune cross-sratification in a turbidite sequence, Cloridorme Formation, Gaspe, Quebec. Sedimentology **17** : 51 - 68.

Skipper, K. and Bhattacharjee, S. L., 1978. Backset bedding in turbidites : a further example from the Cloridorme Formation (Mid-Ordovician), Gaspe, Quebec. J. Sed. Pet. **48** : 193 - 202.

Skipper, K. and Middleton, G. V., 1975. Sedimentary structures and depositional mecnanics of certain Ordovician turbidites, Cloridorme Formation, Gaspe Peninsula, Quebec. Can. J. Earth Sci., **12**: 1934-1952.

Smith, R. K., 1988. Travelling waves and bores in the Lower Atmosphere : The 'Morning Glory' and related phenomena. Earth Sci. Rev.**25** : 267 - 290.

Stacey, M. W. and Bowen, A. J., 1988a. The vertical structure of density and turbidity currents:- Theory and observations. J. Geophys. Res. **93**: 3528-3542.

Stacey, M. W. and Bowen, A. J., 1988b. The vertical structure of turbidity currents and a necessary condition for self-maintenance. J. Geophys. Res. **93**: 3543-3553.

Stow, D. A. V. and Bowen, A. J., 1980. A physical model for the transport and sorting of fine-grained sediments by turbidity currents. Sedimentology **27**: 31-46.

Thorpe,S. A., 1969. Experiments on the instability of stratified shear flows: Immiscible fluids. J. Fluid Mech. **39**: 25-48.

Thorpe, S. A., 1971a Experiments on the instability of stratified shear flows: Miscible fluids. J. Fluid Mech. **46**: 299-319.

Thorpe, S. A., 1971b. Asymmetry of the internal seiche in Loch Ness. Nature **231** : 306 - 308.

Thorpe, S. A., 1973. Experiments on the instability and turbulence in a stratified shear flow. J. Fluid Mech. **61**: 731-751.

Thorpe, S. A., 1978a. On the shape and breaking of finite amplitude internal gravity waves in a shear flow. J. Fluid Mech. **85** : 7 - 31.

Thorpe, S. A., 1978b. On internal gravity waves in an accelerating shear flow. J. Fluid Mech. **88** : 623 - 639.

Thorpe, S. A., Hall, A. and Crofts, I., 1972. The internal surge in Loch Ness. Nature **237** : 96 - 98.

Thorpe, S. A., Hall, A. J. and Hunt, S., 1983. Bouncing internal bores of Ardmucknish Bay, Scotland. Nature **306**: 167-169.

Turner, J. S., 1973. Buoyancy effects in fluids. Cambridge University Press. 367pp.

Van Andel, Tj. H. and Komar, P. D., 1969. Ponded sediments of the Mid-Atlantic Ridge between 22° and 23° North latitude. Bull. geol. Soc. Am. **80** : 1163-1190.

Von Karmen, T., 1940. The engineer grapples with non-linear problems. Bull. Am. Math. Soc. **46**: 615-683.

Walker, L.R., 1973. Interfacial solitary waves in a two fluid medium. Phys. Fluids **16** : 1796 - 1804.

Walker, R. G., 1965. The origin and significance of the internal sedimentary structures in turbidites. Proc. Yorks. geol. Soc. **35**: 1-32.

Walker, R.G., 1978. Deep water sandstone facies and ancient submarine fans: Models for exploration for stratigraphic traps. Am. Ass. Pet. Geol. Bull. **62**: 932-966.

Walton, E.K., 1967. The sequence of internal structures in turbidites. Scott. J. Geol. **3**: 306-317.

Williams, G. P., 1967. Flume experiments on the transport of a coarse sand. Prof. Paper U.S. geol. Surv. **262-B**: 1-31.

Wood, I. R., 1965. Studies in unsteady self preserving turbulent flows. Univ. N. S. Wales, Manly Vale, Australia. Rep. No. **81**: 152pp.

Wood, I. R. and Simpsom, J. E., 1984. Jumps in layered miscible fluids. J. Fluid Mech. **140**: 329-342.

Woodcock, N. H., 1976. Ludlow Series slumps and turbidites. Lond. Geol. Ass. Proc. **87**: 169-182.

Wright, L. D., Wiseman, W. J., Bornhold, B. D., Prior, D. B., Suhayda, J. N., Keller, G.H., Yang, Z.-S. and Fan, Y.B., 1988. Marine dispersal and deposition of Yellow River silts by gravity driven underflows. Nature **232**: 629-632.

Yih, C-S. and Guha, C. R., 1955. Hydraulic jump in a fluid system of two layers. Tellus **7**: 358-366.

Zabusky, N. J. and Kruskal, M. D., 1965. Interaction of "solitons" in a collisionless plasma and the recurrence of initial states. Phys. Rev. Letters **15** : 240 - 243.

SIGNIFICANT ERRORS ENCOUNTERED DURING THE EXPERIMENTS

2A.1 Measurement error

Below is an estimation of the errors associated with the measurement of the flow analysis features of a density current taken from the video monitor during flow analysis.

(i) Time (s)	±0.01sec.
(ii) Height of head (h_1)	±0.6cm
(iii) Horizontal position of nose	±0.1cm
(iv) Thickness of body (h_2) ±0.1cm	
(v) Thickness of bulge of fluid at ramp (D):	±0.3cm
(vi) Flow thickness prior to arrival of reverse current:	±0.1cm
(vii) Horizontal position of reverse flow:	±0.5cm
(viii) Height from base of flume tank to crest of reverse flow waveform:	±0.1cm (Type A bore) ±0.3cm (Type C bore)
(ix) Thickness between waveforms of reverse flow (d_2)	±0.1 (Type A bore) ±0.2cm (Type C bore)

2A.2 The non-reproducability of the experiments.

Individual experiments with the same starting conditions were repeated upto 7 times in order to reduce possible data error and scatter. A significant variation in the data measured from these repeated runs was found to occur due to non-reproducability of individual experiments. Apart from the errors incurred that are listed in 2A.1, the reason for only limited run reproducability was unpredictable although it was possible to calculate typical data ranges as listed below (for the experiments in the CTA series).

(i) Velocity of advance of the head (U_h):	10cm s^{-1} ± 1.1
(ii) Thickness of head (h_1):	3.6cm ± 0.5
(iii) Thickness of body (h_2):	4.7cm ± 0.6

(iv) Velocity of reverse flow (U_b): 5.9cm s^{-1} ±0.4

(v) Thickness from the base of the flume tank
  to the crest of the reverse flow waveform (a_t): 2.0cm ± 0.2cm

As can be seen from the above data, U_h shows a 9% range on the mean in contrast to the measurement errors on the distance and the time values that were used in the calculation of U_h, which are only of the order of 0.7%.

APPENDIX 2B
(overleaf)

EXPERIMENTAL SET-UP CONDITIONS AND DATA

Experiment	Ramp angle (°)	ρ_1	ρ_1/ρ_2	$\Delta\rho$	$\Delta\rho/\rho_1$	U_h	h_1	h_1/H	h_2	h_2/H
S5/1.034	5°	1.034	1.035	0.035	0.034	8.8	7.6	0.506	0.8	0.053
S5/1.068	"	1.068	1.069	0.069	0.065	12.9	7.3	0.487	0.9	0.060
S5/1.100	"	1.100	1.099	0.101	0.092	16.1	7.7	0.513	1.2	0.080
S5/1.128	"	1.128	1.129	0.129	0.114	18.4	6.8	0.453	0.9	0.060
S5/1.152	"	1.152	1.153	0.153	0.133	18.9	7.0	0.467	0.9	0.060
S5/1.178	"	1.178	1.179	0.179	0.152	19.1	7.1	0.473	1.4	0.093
S10/1.012	10°	1.011	1.012	0.012	0.012	6.2	7.0	0.467	1.3	0.087
S10/1.034	"	1.034	1.035	0.035	0.034	8.8	7.5	0.500	1.4	0.093
S10/1.068	"	1.068	1.068	0.069	0.065	12.0	6.5	0.433	0.8	0.053
S10/1.100	"	1.100	1.099	0.101	0.092	15.8	7.4	0.493	1.2	0.080
S10/1.128	"	1.128	1.129	0.129	0.114	17.2	6.9	0.460	0.7	0.047
S10/1.155	"	1.155	1.156	0.156	0.135	18.0	6.3	0.420	1.3	0.087
S10/1.179	"	1.179	1.180	0.180	0.153	21.2	6.8	0.453	1.0	0.067
S15/1.012	15°	1.012	1.013	0.013	0.013	6.0	7.0	0.467	1.0	0.067
S15/1.034	"	1.034	1.035	0.035	0.034	9.1	7.2	0.480	1.0	0.067
S15/1.068	"	1.068	1.069	0.069	0.065	13.9	7.1	0.473	0.9	0.060
S15/1.128	"	1.128	1.129	0.129	0.114	19.6	6.9	0.460	0.9	0.060
S15/1.178	"	1.178	1.179	0.179	0.152	21.7	6.9	0.460	0.9	0.060
S20/1.012	20°	1.012	1.013	0.013	0.013	6.2	7.1	0.473	0.9	0.060
S20/1.068	"	1.068	1.069	0.069	0.065	12.5	6.9	0.460	0.9	0.060
S20/1.128	"	1.128	1.128	0.129	0.114	17.2	6.5	0.433	0.7	0.047
S20/1.178	"	1.178	1.179	0.179	0.152	21.3	6.6	0.440	0.8	0.053

Table 2B.1: Experimental set-up and data for the S5 to S20 series of experiments

Experiment	Fr_H	Fr_{h_2}	U_b wave 1	Bore type A, B or C	d_l	d_0	d_l/d_0	a_t wave 1	d_2	$ub = *$ $f(d_l/d_0)$
S5/1.034	0.394	1.706	4.8	C	1.5	1.0	1.500	1.6	1.1	1.4
S5/1.068	0.417	1.073	7.4	"	1.6	0.9	1.778	1.9	1.2	1.6
S5/1.100	0.439	1.551	7.6	"	1.7	1.2	1.417	1.8	1.5	1.3
S5/1.128	0.448	1.830	9.7	"	1.5	1.0	1.500	1.6	1.3	1.4
S5/1.152	0.427	1.743	9.9	"	1.6	1.2	1.333	1.8	1.3	1.3
S5/1.178	0.404	1.322	10.7	"	1.6	1.3	1.231	1.8	0.7	1.2
S10/1.012	0.457	1.553	4.6	B	1.9	0.7	2.714	2.3	1.3	2.3
S10/1.034	0.394	1.291	7.3	"	2.2	0.8	2.750	2.2	1.7	2.3
S10/1.068	0.388	1.680	10.0	"	2.5	1.0	2.500	2.5	2.2	2.1
S10/1.100	0.431	1.522	11.8	"	2.7	1.0	2.700	2.9	2.4	2.3
S10/1.128	0.419	1.937	12.0	"	2.3	1.0	2.300	2.4	1.7	2.0
S10/1.155	0.404	1.373	12.7	"	2.6	1.1	2.364	2.8	1.9	2.0
S10/1.179	0.448	1.734	15.1	"	2.6	1.0	2.600	2.7	2.0	2.2
S15/1.012	0.442	1.713	6.2	A	2.8	0.6	4.667	3.5	2.5	3.7
S15/1.034	0.408	1.580	8.8	"	2.4	0.7	3.429	3.3	1.5	2.7
S15/1.068	0.450	1.839	10.8	"	2.6	0.8	3.250	3.3	1.8	2.7
S15/1.128	0.478	1.951	15.9	"	3.0	0.9	3.333	3.6	2.7	2.7
S15/1.178	0.458	1.871	18.1	"	3.0	1.0	3.000	3.3	2.2	2.5
S20/1.012	0.457	1.866	6.2	"	2.9	0.9	3.222	3.6	2.2	2.3
S20/1.068	0.405	1.654	10.5	"	3.3	0.9	3.667	3.9	2.6	3.0
S20/1.128	0.419	1.941	15.5	"	3.3	0.9	3.667	4.1	2.2	3.0
S20/1.178	0.450	1.948	21.3	"	4.0	0.9	4.444	4.6	3.6	3.5

Other information

(i) $\varrho_2 = 0.999$g cm^{-3}

(ii) Total flow depth = 15.0cm

(iii) Foot of ramp from end wall of lock = 270.5cm

(iv) Centre of field of view of video = 177.5cm

* $\quad U_b = g'd_0[0.5(d_1/d_0)(1 + d_1/d_0)]^{0.5}$

Experiment	ub/(g'd0)	U_b/U_h
S5/1.034	0.832	0.545
S5/1.068	0.977	0.574
S5/1.100	0.732	0.472
S5/1.128	0.915	0.527
S5/1.152	0.791	0.523
S5/1.178	0.768	0.560
S10/1.012	1.570	0.742
S10/1.034	1.416	0.830
S10/1.068	1.252	0.833
S10/1.100	1.246	0.747
S10/1.128	1.131	0.698
S10/1.155	1.053	0.706
S10/1.179	1.235	0.712
S15/1.012	2.286	1.033
S15/1.034	1.825	0.967
S15/1.068	1.516	0.777
S15/1.128	1.583	0.811
S15/1.178	1.481	0.834
S20/1.012	1.866	1.000
S20/1.068	1.389	0.840
S20/1.128	1.543	0.901
S20/1.178	1.837	1.000

Experiment	Ramp Angle (°)	Video position* (cm) (i)	(ii)	ρ_1	ρ_1/ρ_2	$\Delta\rho$	$\Delta\rho/\rho_1$	U_h	h_1
FE5/1.012	5	50.5	162	1.012	1.013	0.013	0.013	–	–
FE5/1.068	"	"	"	1.068	1.069	0.069	0.065	9.1	5.1
FE5/1.128	"	"	"	1.128	1.129	0.129	0.114	13.5	5.6
FE5/1.178	"	"	"	1.178	1.179	0.179	0.152	–	–
FE10/1.012	10	"	"	1.012	1.013	0.013	0.013	–	–
FE10/1.068	"	"	"	1.068	1.069	0.069	0.065	8.7	6.0
FE10/1.128	"	"	"	1.128	1.129	0.129	0.114	13.1	5.4
FE10/1.178	"	"	"	1.178	1.179	0.179	0.152	13.7	5.4
FE15/1.012	15	"	"	1.012	1.013	0.013	0.013	–	–
FE15/1.068	"	"	"	1.068	1.069	0.069	0.065	8.8	6.5
FE15/21.128	"	"	"	1.128	1.129	0.129	0.114	14.1	6.1
FE15/1.178	"	"	"	1.178	1.179	0.179	0.152	14.2	5.5
FE20/1.012	20	"	"	1.012	1.013	0.013	0.013	–	–
FE20/1.068	"	"	"	1.068	1.069	0.069	0.065	8.8	6.1
FE20/1.128	"	"	"	1.128	1.129	0.129	0.114	13.5	5.8
FE20/1.178	"	"	"	1.178	1.179	0.179	0.152	17.0	6.1

* Position of centre of field of view from end wall of lock.

Table 2.B.2: Data for the FE5 to FE20 series of experiments

Experiment	h₁/H	h₂	h₂/H	Fr_H	D	D/H	U_b (wave 1)	Bore type A, B or C	d_I
FE5/1.012	–	–	–	–	–	–	–	–	–
FE5/1.068	0.34	1.6	0.107	0.506	–	–	–	–	–
FE5/1.128	0.373	1.3	0.087	0.347	–	–	–	–	–
FE5/1.178	–	–	–	–	–	–	–	–	–
FE10/1.012	–	–	–	–	–	–	3.9	C	1.8
FE10/1.068	0.4	1.5	0.1	0.446	1.1	0.073	9.4	"	1.9
FE10/1.128	0.36	1.6	0.107	0.532	1.1	0.073	13.0	"	2.4
FE10/1.178	0.36	1.5	0.1	0.482	1.7	0.113	15.4	"	2.4
FE15/1.012	–	–	–	–	–	–	4.7	"	2.0
FE15/1.068	0.433	1.9	0.127	0.433	1.7	0.113	10.7	"	2.5
FE15/21.128	0.406	1.5	0.1	0.539	2.2	0.145	14.9	"	2.6
FE15/1.178	0.367	1.4	0.093	0.495	2.5	0.167	18.1	B	2.7
FE20/1.012	–	–	–	–	–	–	14.5	C	1.7
FE20/1.068	0.407	1.7	0.113	0.447	1.7	0.113	10.4	"	2.6
FE20/1.128	0.387	2.1	0.14	0.529	2.2	0.145	14.4	"	2.6
FE20/1.178	0.407	2.0	0.133	0.563	2.3	0.153	18.6	B	2.9

Experiment	do	d₁/do	at wave 1	d₂	L	L/H	$U_b = **$ f(d1/d0)	$U_b/(g'd_0)$	U_b/U_k
FE5/1.012	-	-	-	-	-	-	-	-	-
FE5/1.068	1.4	-	2.0	-	-	-	-	-	-
FE5/1.128	1.1	-	-	-	-	-	-	-	-
FE5/1.178	-	-	-	-	-	-	-	-	-
FE10/1.012	1.1	1.636	2.1	1.4	5.1	0.340	1.4	1.062	-
FE10/1.068	1.2	1.583	2.3	1.2	4.7	0.313	1.6	1.077	1.081
FE10/1.128	1.6	1.500	2.9	1.7	4.1	0.273	1.2	0.971	0.992
FE10/1.178	1.3	1.846	2.8	1.6	3.8	0.253	1.7	1.105	1.124
FE15/1.012	1.2	1.667	2.7	1.3	4.5	0.300	1.6	1.225	-
FE15/1.068	1.3	1.923	3.3	1.4	4.8	0.320	1.7	1.178	1.216
FE15/21.128	1.3	2.000	3.2	1.4	4.7	0.313	1.7	1.234	1.057
FE15/1.178	1.4	1.929	3.6	1.9	5.9	0.393	1.7	1.252	1.275
FE20/1.012	0.8	2.125	2.2	1.1	5.0	0.333	1.9	1.437	-
FE20/1.068	1.2	2.167	3.8	1.7	3.8	0.253	1.9	1.192	1.182
FE20/1.128	1.4	1.857	3.3	1.7	5.7	0.380	1.7	1.149	1.067
FE20/1.178	1.6	1.813	3.6	1.9	7.3	0.487	1.6	1.203	1.094

Other information:

(i) $\rho_2 = 0.999 \text{g cm}^{-3}$

(ii) Total flow depth = 15.0cm

(iii) Foot of ramp from end wall of lock = 370.5cm

** $U_b = g'd_0[0.5(d_1/d_0)(1 + d_1/d_0)]^{0.5}$

Experiment	Ramp angle (°)	ρ_1	ρ_2	ρ_1/ρ_2	$\Delta\rho$	$\Delta\rho/\rho_1$	H	U_b	h_1	h_1/H
CD5/5	5	1.067	0.998	1.068	0.069	0.065	5.0	6.1	1.7	0.340
CD5/7.5	"	1.068	0.999	1.069	"	"	7.5	8.2	3.2	0.427
CD5/10	"	"	"	"	"	"	10.0	8.9	4.3	0.043
CD5/12.5	"	"	"	"	"	"	12.5	11.2	6.3	0.504
CD5/15	"	"	"	"	"	"	15.0	12.9	7.3	0.487
CD10/5	10	"	"	"	"	"	5.0	5.8	1.8	0.360
CD10/7.5	"	"	"	"	"	"	7.5	8.4	3.4	0.453
CD10/10	"	"	"	"	"	"	10.0	10.1	4.7	0.470
CD10/12.5	"	"	"	"	"	"	12.5	11.4	6.0	0.480
CD10/15	"	"	"	"	"	"	15.0	12.0	6.5	0.433

Table 2.B3

Experimental set-up and data for the CD5 to CD10 series of experiments.

Experiment	h$_z$	h$_2$/H	Fr$_H$	Fi h$_z$	U$_b$ wave 1	Bore type A, B or C	d$_1$	d$_o$	d$_1$/d$_o$	d$_1$/H
CD5/5	1.1	0.220	0.342	0.729	-	-	-	-	-	-
CD5/7.5	1.2	0.160	0.375	0.938	-	-	-	-	-	0.120
CD5/10	1.7	0.170	0.353	0.856	-	-	-	-	-	0.100
CD5/12.5	0.8	0.064	0.397	1.570	-	-	-	-	-	0.072
CD5/15	0.9	0.060	0.417	1.703	-	-	-	-	-	0.060
CD10/5	1.1	0.220	0.325	0.694	-	-	-	-	-	-
CD10/7.5	1.1	0.147	0.385	1.005	8.4	C	1.6	1.2	1.304	0.160
CD10/10	1.7	0.170	0.401	0.973	7.5	"	1.7	1.0	1.833	0.100
CD10/12.5	0.8	0.064	0.405	1.600	8.8	B	2.2	0.8	2.750	0.064
CD10/15	0.8	0.053	0.388	1.680	10.0	"	2.5	1.0	2.500	0.067

Experiment	d_i/H	U_b/U_{fi}
CD5/5	-	-
CD5/7.5	-	-
CD5/10	-	-
CD5/12.5	-	-
CD5/15	-	-
CD10/5	-	-
CD10/7.5	0.211	1.000
CD10/10	0.174	0.743
CD10/12.5	0.176	0.779
CD10/15	0.167	0.833

Other information:

(i) Centre of field of view of video
from end wall of lock = 177.5cm

(ii) Foot of ramp from end wall of lock = 270.5cm

Experiment	Ramp angle (°)	ρ_1	$\Delta\rho$	$\Delta\rho/\rho_1$	U_h	U_b wave 1	Bore type A, B or C	U_b/U_h
T5/1.178	5	1.178	0.178	0.151	12.0	5.3	s	0.442
T5/1.128	"	1.128	0.128	0.114	9.7	4.2	"	0.429
T5/1.069	"	1.069	0.069	0.064	7.0	3.1	"	0.442
T10/1.178	10	1.178	0.178	0.151	12.2	5	"	0.408
T10/1.128	"	1.128	0.128	0.114	10.0	4.8	"	0.479
T10/1.069	"	1.069	0.069	0.064	7.1	3.4	"	0.478
T20/1.178	20	1.178	0.178	0.151	12.0	6.6	"	0.547
T20/1.128	"	1.128	0.128	0.114	9.8	6.1	"	0.625
T20/1.069	"	1.069	0.069	0.064	7.1	4.1	"	0.572

Table 2A.4: Data for the T5 to T20 series of experiments

Other information

(i) $\rho_\lambda = 1.000 \text{g cm}^{-3}$

(ii) Total flow depth = 10cm

(iii) Video positioned directly overhead

(iv) Strike of ramp at $45°$ to channel axis

Name	Ramp angle	Video position (cm)*
PP5/102.5	5°	102.5
PP5/209.5	"	209.5
PP10/102.5	10°	102.5
PP10/209.5	"	209.5
PP15/102.5	15°	102.5
PP15/209.5	"	209.5
PP20/102.5	20°	102.5
PP20/209.5	"	209.5
SC 1	8°	209.5
CTA	"	209.5

(i) $\rho_1 = 1.053$ g.cm-3
(ii) $\rho_2 = 0.999$ g.cm-3
(iii) Foot of ramp from end-wall of lock = 288.5cm

* Centre of field of view from end wall of lock

Table 2 B.5
Experimental set-up for the series PP1 to PP8; SC1 and the CTA velocity profiles.

Name	Sediment	ϱ, (g cm-3)	Ramp angle	Ramp position*	Total flow depth	Video position**
Sed 1	Plaster of Paris	1.131	10°	270.5	10	
Sed 2	Silica flour	1.068	20°	370.5	15	50.5; 162.0

*   Position of foot of ramp from end-wall of lock.

** Position of centre of field of view from end-wall of lock.

Table 2.B 6

Experimental set-up for the sediment suspension experiments

CONSTANT TEMPERATURE ANEMOMETRY

(i) Equipment supplied by Dantec Electronics Ltd., Techno House, Redcliffe Way, Bristol, BS1 6NU.

(ii) Technical details:

(a) A steel clad probe was used in conjunction with a CTA Bridge (type 56C17) plugged into a 56C01 anemometer and a 56B10 mainframe control box.

(b) A steel clad probe was selected for use because it is robust and is capable of withstanding the impact of particles in the flow that may collide with it. It is the recommended probe type for the accurate measurement of mean flow velocities and is also good for the determination of velocity fluctuations. Unfortunately, the omnidirectional nature of the probes data collection does not allow the distinction of different flow directions to be distinguished or calculated.

(c) Characteristics of the steel clad probe include:
   1) A nickel sensor,
   2) Sensor dimensions: 1.1mm (diameter) x 6mm (length),
   3) Resistance at 10°C is 10Ω,
   4) Minimum measured velocity in water: 0.005m s^{-1},
   5) Maximum measured velocity in water: 3m s^{-1} and,
   6) Frequency respose: 10 to 15 Hz.

(iii) Problems encountered during its use.

During the couse of this present study, certain technical problems were encountered with the CTA equipment. Firstly, it was not possible to calibrate the probes accurately and, as a result, the velocity profiles that were obtained were only of a semi-quantitative nature. Secondly, set-up problems with the calibration section of Dantec's newly introduced CTA software package ("Acqwire") so limiting the CTA's value during this study.

Lecture Notes in Earth Sciences